FOCUS
REVIEW GUIDE

to accompany
Raven *Biology* AP® Edition

Mc
Graw
Hill
Education

About the AP Consultant

Darrel James received his BS in Biology from Pacific University in Oregon and his MS in Marine Science from Oregon State. He taught biology for thirty years, and is presently teaching at Beyer High School in Modesto, CA. Darrel teaches Pre-AP as well as AP Biology and is the department chairperson. Darrel has served as a reader, table leader, and assistant chief reader for the AP Biology exam since 1990. He has led numerous one-day AP Biology Curriculum and grading workshops for the College Board, and he has presented week-long institutes for the past twenty-four summers. Darrel is a member of the California Commission on Science and Technology. He is the vice-chair of the California Teachers Advisory Council, an arm of the CCST which advises the state on STEM and digitally enhanced education. Darrel has led marine biology trips to Hawaii for his AP Biology students. He is also an avid cyclist. He loves to teach and show people how much fun biology can be.

COVER: (tl)mike black photography/Getty Images, (tr)cherokee4/iStock /Getty Images Plus/Getty Images, (b)LAGUNA DESIGN/Science Photo Library/Getty Images.

MHEonline.com

Send all inquiries to:
McGraw-Hill Education
8787 Orion Place
Columbus, OH 43240

ISBN: 978-0-07-667254-7
MHID: 0-07-667254-9

Printed in the United States of America.

3 4 5 6 7 8 QVS 23 22 21 20 19

Table of Contents

*Recall It/Review It sections contain material that is not specifically part of the AP curriculum, but students should have a basic understanding of this content.

*Recall It/Review It sections contain material that is not specifically part of the AP curriculum, but students should have a basic understanding of this content.

*Recall It/Review It sections contain material that is not specifically part of the AP curriculum, but students should have a basic understanding of this content.

*Recall It/Review It sections contain material that is not specifically part of the AP curriculum, but students should have a basic understanding of this content.

Note-Taking Tips

Your notes are a reminder of what you learned in class. Taking good notes can help you succeed in science. The following tips will help you take better classroom notes.

- Before class, ask what your teacher will be discussing in class. Review mentally what you already know about the concept.
- Be an active listener. Focus on what your teacher is saying. Listen for important concepts. Pay attention to words, examples, and/or diagrams your teacher emphasizes.
- Write your notes as clearly and concisely as possible. The following symbols and abbreviations may be helpful in your note-taking.

Word or Phrase	Symbol or Abbreviation	Word or Phrase	Symbol or Abbreviation
for example	e.g.	and	+
such as	i.e.	approximately	≈
with	w/	therefore	∴
without	w/o	versus	vs

- Use a symbol such as a star ★ or an asterisk * to emphasize important concepts. Place a question mark ? next to anything that you do not understand.
- Ask questions and participate in class discussion.
- Draw and label pictures or diagrams to help clarify a concept.
- When working out an example, write what you are doing to solve the problem next to each step. Be sure to use your own words.
- Review your notes as soon as possible after class. During this time, organize and summarize new concepts and clarify misunderstandings.

Note-Taking Don'ts

- **Don't** write every word. Concentrate on the main ideas and concepts.
- **Don't** use someone else's notes. They may not make sense.
- **Don't** doodle. It distracts you from listening actively.
- **Don't** lose focus or you will become lost in your note-taking.

Using Your FOCUS REVIEW GUIDE

This review guide was developed with the AP student in mind. The activities within each chapter will help you to focus on and review the key content in the chapter as it relates to the AP Biology Curriculum.

The **AP Essential Knowledge** charts review the parts of the AP Curriculum covered in the chapter. You will find that these match the Essential Knowledge charts in your student edition.

The **Chapter Overview** summarizes the chapter pinpointing the parts of the AP Curriculum covered in the chapter.

The **Recall It** summaries will help you review prerequisite knowledge necessary to understand the key concepts covered in the chapter.

Review It activities will help you to review key concepts that you should have mastered in your study of the chapter.

Different visual organizers help you to analyze and summarize information and remember content.

Use It activities allow you to apply what you've learned in the chapter.

This section notes the **Essential Knowledge** covered in the section. The activities in the **Use It** section correlate to these standards.

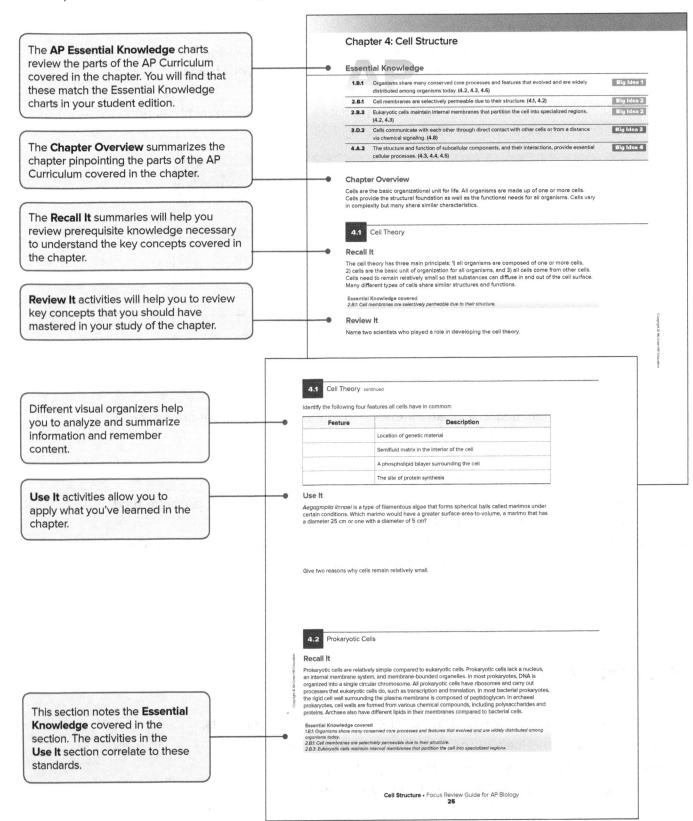

Sections that are not covered by the AP Curriculum contain only a **Recall It** summary and **Review It** activities. While these sections do not present material directly covered on the AP Exam, they do provide important content AP Biology students should know and be able to recall and apply.

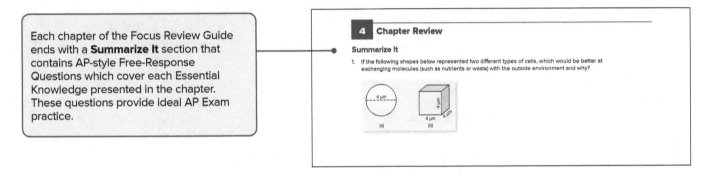

At the end of each Focus Review Guide Chapter, there is a Summarize It section. This section provides an excellent review of the chapter's AP content.

Each chapter of the Focus Review Guide ends with a **Summarize It** section that contains AP-style Free-Response Questions which cover each Essential Knowledge presented in the chapter. These questions provide ideal AP Exam practice.

Some chapters are not covered by the AP Curriculum, but include illustrative examples that can be used to illustrate particular concepts. These chapters are classified as Extending Knowledge chapters. For these chapters we've devised a special review that allows you to take what you've learned and apply it to these illustrative examples.

The **AP Extending Knowledge** section helps to illustrate how the chapter relates to AP content.

The **Review It** activities in Extending Knowledge chapters focus on reviewing how the chapter content relates to illustrative examples of AP Biology Curriculum.

The **Summarize It** section applies the illustrative example content to an AP-style Free-Response Question. These are similar to one you might find on the AP Biology Exam.

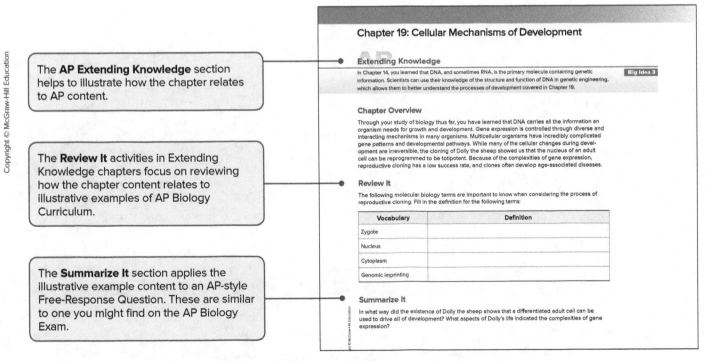

Chapter 1: The Science of AP Biology

Essential Knowledge

Big Idea 1	The process of evolution drives the diversity and unity of life. **(1.1, 1.6)**
Big Idea 2	Biological systems utilize free energy and molecular building blocks to grow, to reproduce and to maintain dynamic homeostasis. **(1.2)**
Big Idea 3	Living systems store, retrieve, transmit and respond to information essential to life processes. **(1.3)**
Big Idea 4	Biological systems interact, and these systems and their interactions possess complex properties. **(1.4)**

Chapter Overview

There are four Big Ideas that you will explore while studying AP Biology: (1) The process of evolution drives diversity and unity of life, (2) Biological systems use energy to grow, reproduce, and maintain homeostasis, (3) Living systems store, retrieve, transmit and respond to information in the environment, and (4) Biological systems are complex and are composed of parts with interact with each other. This chapter provides an overview of these different ideas, and why these ideas are fundamental to the study of biology. This chapter also introduces the seven science practices that will guide your study of AP Biology.

1.1   Big Idea 1: Evolution

Recall It

Life has existed on this planet for billions of years. During this vast time span, organisms have changed genetically and physically, many have gone extinct, and new species have emerged. Evolution is the change in species over time, and evolution often occurs through the process of natural selection. This now fundamental theory in biology was first published in 1859 by English naturalist Charles Darwin. Subsequent findings from many disciplines support the theory of evolution, including: physics, geology, molecular biology, mathematics, genetics, anatomy, and proteomics. Biological evolution lies at the center of AP Biology.

Review It

Determine whether or not the following structures are homologous **(H)** or analogous **(A)**:

The wing of a butterfly and the wing of a bat

The bones in the forelimb of a cat and the bones in the wing of a bat

Number of amino acids in dog hemoglobin and number of amino acids in frog hemoglobin

The spines of a cactus and the spines of a hedgehog

Provide an example of how the following disciplines provide evidence for biological evolution.

Discipline	Evidence
Geology	
Physics	
Genetics	
Mathematics	

How did Darwin use Thomas Malthus' work to describe changes in animal population?

Today, many household have pet dogs and cats. Humans have not always had pets, though. Where did these pets come from?

1.2 Big Idea 2: Energy and Molecular Building Blocks

Recall It

Energy flows through living systems, driving all biological processes. Organisms require energy to grow, reproduce, and maintain organization. Living things have developed complex mechanisms for energy capture and storage. Living systems are highly coordinated from regulations at the cellular level to timing and coordination within ecosystems.

Review It

Where does all energy needed by life on Earth come from?

Define *homeostasis*.

Membranes are said to be "key" to all cellular things. List two reasons why.

1.3 Big Idea 3: Information Storage, Transmission, and Response

Recall It

Information for life is stored, transmitted, and responded to at many levels. At the genetic level, deoxyribonucleic acid (DNA) stores all the information needed by cells to maintain structure and function. DNA needs to be copied and passed on to future generations for life to continuously operate. Genetic variation leads to new traits in different organisms. There are many ways cells and organisms transmit messages and respond to incoming information: cells send signals to other cells, birds may sing out a warning song to alarm other birds of a predator, and dropping temperatures may tell a tree that it's time to drop its leaves and become dormant.

Review It

Where can the "internal blueprint" of a cell be found?

Describe three ways in which the cell cycle allows life to continue.

List two ways genetic variation may occur.

How do you communicate an idea you've had to a friend, and how does your friend receive that message?

Recall It

There is a saying that "the whole is greater than the sum of its parts," and that is certainly true for living systems. Cooperation takes place at multiple levels in living system, from the proteins in a cell to populations in an ecosystem. Damaging or destroying any part of a living system will affect the whole as the parts are all tightly interconnected. Any place there can be more diversity or more options in a living system, the better change life has of survival.

Review It

Explain four systems in your body that you depend on for survival.

Describe the difference between cooperation and competition.

Imagine there was a farm that grew only one strain of corn (Corn Y). One summer, a swarm of insects descended on the area that only ate Corn Y. All of the crops were destroyed. What could the farmers done before the growing season to decreased the risk of losing all of their crops?

1.5 AP Science Practices

Recall It

There are seven AP Science Practices that you will be required to know and practice. Science is descriptive, so you must be able to explain ideas and results through models, diagrams, and graphs. It is also expected that you are able to use math appropriately. You will learn how develop a good scientific question; one that can be researched and lead to credible answers. Anyone who studies science must be able to collect, analyze, and evaluate data, as well as provide justification for any claims. Finally, it is important to connect what you learn in biology to other scientific disciplines.

Review It

Provide a definition for the following:

Scientific Word	Definition
Theory	
Deductive reasoning	
Inductive reasoning	

Practice the Seven AP Science Principals.

1. Communicate with a model	Draw a diagram of where you sit in your classroom.	
2. Use math appropriately	Calculate the area of a 2 × 2 m square.	
3. Question scientifically	Ask a question that can be researched and answered.	
4. Collect data responsibly	Determine the colors of your family members eyes.	
5. Analyze and evaluate data	Create a bar chart showing the data you collected above.	
6. Justify your conclusions	Three friends all ate at a picnic. Two ate the potato salad and became sick. The other friend did not eat the potato salad and did not get sick. What made the two friends sick?	
7. Expanding understanding and connections	Some plants have the ability to take up and store hazardous metals in their roots. How could this be used to help clean up waste around an old mining operation?	

Recall It

The diversity we see in living organisms is astounding and has evolved over billions of years from one common ancestor. Living organisms are grouped into three domains, comprised of six kingdoms based on their differences. Living organisms have seven characteristics in common: 1) all are composed of one or more cells, 2) they respond to stimuli, 3) they are highly complex and ordered, 4) they grow, reproduce, and transmit genetic information to their offspring, 5) they requires energy, 6) they can maintain homeostasis, and 7) they are capable of adapting to the environment. Living systems are organized from atoms all the way to the biosphere, and, at each level, emergent properties appear.

Essential Knowledge covered
1.D.2: Scientific evidence from many disciplines supports models of the origin of life.

Review It

Provide one piece of evidence that biologist point to that supports the idea that all living organisms today may have descended from a simple cellular creature 3.5 BYA.

List and explain the layers of hierarchical organization of living systems

Using the seven characteristics shared by living systems, justify that you are alive.

Chapter 2: The Nature of Molecules and the Properties of Water

Essential Knowledge

2.A.3   Organisms must exchange matter with the environment to grow, reproduce and maintain organization. **(2.4, 2.5)**   Big Idea 2

Chapter Overview

Cells and organisms are made up of atoms that have been linked together to form molecules, all of which are governed by the laws of physics. Molecules are formed from atoms through different chemical interactions. The chemical properties which make water molecules unique also make life possible.

2.1   The Nature of Atoms

Recall It

Living and non-living things are made up of atoms. Atoms have a nucleus which contains protons and neutrons (except for hydrogen which has no neutrons). Electrons surround the nucleus in what are known as orbitals. The number of electrons an atom has determines the atom's chemical behaviors. Atoms with the same number of protons have the same chemical properties and belong to the same element. Isotopes are atoms of a single element that possess a different number of neutrons. Unstable isotopes may lead to radioactive decay.

Review It

Given the definition on the left, fill in the correct vocabulary word on the right:

Definition	Vocabulary Word
The structure of electrons within the atom correspond to quanta, also known as these	
The loss of electrons during a chemical reaction	
The time it takes for one-half of the atoms in a sample to decay	
An atom carrying a negative charge	
When oxidation and reduction occur simultaneously	
An atom carrying a positive charge	

List three of the most abundant isotopes of carbon.

Determine the charge of the following subatomic particles:

Proton	Electron	Neutron

2.2 Elements Found in Living Systems

Recall It

The periodic table displays all of the elements which occur on Earth, 90 of which occur naturally. Elements are arranged by the number of valance electrons (the number of electrons in their outer-most energy level). Elements that have established outermost energy levels which are completely full of electrons are inert or nonreactive.

Review It

List four of the 12 elements which are found in abundance (more than 0.01%) in living systems:

Which orbital will fill first with electrons?

Describe the number of electrons, protons, and neutrons in the following element:

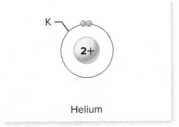

Helium

Is the element above an inert or highly reactive element?

Why is iodine considered a trace element?

2.3 The Nature of Chemical Bonds

Recall It

Chemical bonds join atoms to form molecules. Molecules of more than one element are called compounds. There are several different types of bonds and interactions that occur between atoms to form molecules and compounds. Ionic bonds form when atoms with opposite charges attract. Covalent bonds form when atoms share one or more pairs of electrons. Polar and nonpolar covalent bonds depend on the electronegativity of the atoms in the bond. Bonds are formed and broken through chemical reactions. There are three important factors which influence chemical reactions: temperature, concentration of reactant and products, and catalysts.

Review It

Determine the following type of bond:

Bond Description	Bond
Two atoms share electrons but unequally; there is an area of partial negative charge near one atom and a region of partial positive charge near the second atom	
Electrostatic attraction of oppositely charged ions leads to the formation of a bond	
The strongest covalent bond	

Describe why the following three factors are important to a chemical reaction:

Factor	Function
Temperature	
Concentration of reactant	
Catalyst	

Recall It

Water composed of oxygen and hydrogen joined with highly polar covalent bonds. Water's polarity underlies its chemistry and the chemistry of life. The partial charges formed at both the oxygen and hydrogen ends of the molecule allow for chemical associations known as hydrogen bonding. These bonds are weaker than covalent bonds, but the sheer abundance of these bonds allow for water's many unique physical properties. Water molecules are cohesive, or attracted to other water molecules. Cohesion helps to facilitate the movement and exchange of water between an organism and its environment. Water is also adhesive, creating hydrogen bonds with other polar substance. Adhesion between water and other polar substances allows for phenomena such as capillary action.

Essential Knowledge covered
2.A.3: Organisms must exchange matter with the environment to grow, reproduce, and maintain organization.

Review It

Determine whether or not the following statements are true **(T)** or false **(F)** regarding hydrogen bonding:

Hydrogen bonds push water molecules apart.

Water adheres to nonpolar molecules.

Hydrogen bonds are responsible for the high surface tension of water.

Water carries two partial positive charges near the oxygen atom and two partial negative charges on each hydrogen atom.

Hydrogen bonding is essential for the movement of water in plants.

Decide whether or not **(Y/N)** water can form hydrogen bonds with the following substances:

Water

Oil

Glass

Identify the property of water responsible for the following occurrences:

Property	Occurrence
	A spider walks across a lake.
	Water moves up a small glass tube in a beaker of water.
	Tree leaves pull water up from the roots.

Use It

Draw a ball and stick model of a water molecule. Be sure to label all atoms. Identify the partial charges on each end of the molecule.

Describe how water's polarity allows for the properties of cohesion and adhesion.

Look at the figure to the left. Why does the water rise higher in the tube with the smallest diameter?

2.5 Properties of Water

Recall It

Life is made possible by the many unique properties of water. Organisms and systems maintain constant temperatures through water's high specific heat and high heat of vaporization. Water heats up slowly and holds its temperature longer than almost any other compound, helping to maintain internal temperatures. If organisms become too hot, they can release excess heat through evaporative cooling, such as sweating. Water in solid form is less dense than in its liquid state, so floating ice keeps organisms living in water below from freezing. Water is a solvent, which helps facilitate movement of substances across a cell membrane. Water can also organize nonpolar substances, as nonpolar substances aggregate when interacting with water. Water disassociates to form ions, H^+ and OH^-, which are essential for many organisms to maintain pH levels within their bodies.

Essential Knowledge covered
2.A.3: Organisms must exchange matter with the environment to grow, reproduce, and maintain organization.

Review It

Provide an explanation for how hydrogen bonds allow for the following property of water:

Property	Explanation
Lower density of ice	
Solubility	
High heat of vaporization	
High specific heat	

Use It

Explain two factors that allow water to have so many unique properties.

Explain why a teaspoon of sugar will dissolve in a cup of water but olive oil will float on the surface.

On a hot summer day, you may go for a swim in a pool. Describe in terms of hydrogen bonding how this will make you feel cooler.

A great deal of heat is released by the chemical reactions occurring inside your cells. How are your cells able to survive the countless chemical reactions that occur without being destroyed by all that excess heat?

2.6 Acids and Bases

Recall It

The concentration of hydrogen ions in a solution determines a solution's pH. Acids have a pH lower than 7, and bases have a pH higher than 7. Pure water has a pH of 7 and is called neutral. Buffers are substances which resist the change of pH in a solution. Buffers release hydrogen ions when a base is added to a solution and absorb hydrogen ions when an acid is added. Buffers are very important in maintaining constant pH in biological systems.

Review It

Describe what makes a chemical an acid or a base.

Identify if these solutions have more hydrogen ions, more hydroxide ions, or an equal number of hydrogen and hydroxide ions based on their pH.

Solution	pH	Ion
Coffee	5	
Tears	7.0	
Soda	3.0	
Bleach	12.0	

Human blood has a pH of around 7.4. This pH is kept constant by a chemical known as carbonic acid. If too much H^+ is in the blood, carbonic acid is formed. If a basic substance removes H^+ from the blood, carbonic acid dissociates and releases more H^+. What is this chemical known as?

2 Chapter Review

Summarize It

Using a diagram, describe how the properties of cohesion and adhesion would allow for water to be moved from the ground, through the roots of a plant, up into the leaves.

Chapter 3: The Chemical Building Blocks of Life

Essential Knowledge

2.A.3	Organisms must exchange matter with the environment to grow, reproduce and maintain organization. **(3.1)**	**Big Idea 2**
2.B.1	Cell membranes are selectively permeable due to their structure. **(3.5)**	**Big Idea 2**
3.A.1	DNA, and in some cases RNA, is the primary source of heritable information. **(3.3)**	**Big Idea 3**
4.A.1	The subcomponents of biological molecules and their sequence determine the properties of that molecule. **(3.1, 3.2, 3.3, 3.4, 3.5)**	**Big Idea 4**
4.B.1	Interactions between molecules affect their structure and function. **(3.4)**	**Big Idea 4**

Chapter Overview

The major classes of biological molecules are carbohydrates, nucleic acids, proteins, and lipids. These molecules are the building blocks of life. These molecules allow for all of life's function's including energy storage, data storage, and all biological processes. Biological molecules also maintain the essential structures of living organism, including cell membranes.

3.1 Carbon: The Framework of Biological Molecules

Recall It

All important biological molecules contain carbon. Molecular groups attached to a carbon-hydrogen skeleton, known as functional groups, account for differences in chemical properties. Some molecules, known as isomers, have the same molecular formula but different structures. Changes in the structure of a molecule can affect biological function. Biological macromolecules such as carbohydrates, nucleic acids, proteins, and lipids allow organisms to grow, reproduce, and maintain organization. The long chains of similar subunits in macromolecules are joined by dehydration reactions and are broken down by hydrolysis reactions.

Essential Knowledge covered
2.A.3: Organisms must exchange matter with the environment to grow, reproduce and maintain organization.
4.A.1: The subcomponents of biological molecules and their sequence determine the properties of that molecule.

Review It

Provide a functional group that may be found in the following biological molecules:

Biological Molecule	Functional Group
Carbohydrates	
Lipids	
Nucleic Acids	
Proteins	

Identify these biological macromolecules based on the information about their subunits and function.

Macromolecule	Subunits	Function
	Nucleotides	Encodes genes
	Glucose	Structural support in plant walls
	Glycerol and three fatty acids	Stores energy

Use It

Describe two ways carbon is used by all organisms.

Draw the chiral isomer of the following amino acid:

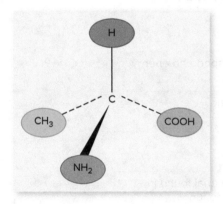

Describe how macromolecules are made and broken apart. How is water involved?

Recall It

Carbohydrates make excellent energy storage molecules because of the abundant carbon-hydrogen bonds they have. Carbohydrates are found in many forms: from monosaccharides (or "simple sugars") typically arranged in ring form, to two linked monosaccharides called disaccharides, to much longer polysaccharide chains. Glucose is the most important six-carbon monosaccharide for energy storage. The empirical formula for glucose, $C_6H_{12}O_6$, is shared by many isomers and stereoisomers, including fructose and galactose. The structural changes between these molecules account for the difference in their properties. Glucose is often transformed into a disaccharide with itself or other monosaccharides in order to facilitate transport throughout an organism, as well as to create important energy sources such as the milk sugar lactose. Likewise, the structure of some polysaccharides allow for better energy storage, such the branched polymers of α-glucose found in starch, or are better for structural support, as is the case in unbranched chains of β-glucose found in the cellulose of plants.

Essential Knowledge covered
4.A.1: The subcomponents of biological molecules and their sequence determine the properties of that molecule.

Review It

List either the basic structure or function of the following carbohydrates:

Carbohydrate	Structure	Function
Glycogen	Long branched chains of α-glucose	
Sucrose		Glucose transport and energy storage
Amylose		Energy storage
Glucose	Six-carbon monosaccharide with a carbonyl group; can be linear or ring-shaped	
Chitin		Structural material found in arthropods and fungi
Ribose	Monosaccharide, five-carbon ring with carbonyl group	
Cellulose		Structural

Use It

When glucose forms a ring in solution, two ring closures can result in two different molecules: α-glucose or β-glucose. If several hundred α-glucose are linked together, what type of resulting polysaccharide would you find?

Why are cows able to break down the β-(1→4) linkages found in cellulose while humans cannot?

3.3   Nucleic Acids: Information Molecules

Recall It

DNA and RNA are the two types of nucleic acids which encode all the information needed for life. DNA stores genetic information, while RNA carries the information from DNA, facilitates protein synthesis, and may even be involved in gene regulation. Nucleic acids are composed of nucleotides. Each nucleotide has a five carbon sugar (ribose in RNA or deoxyribose in DNA), a phosphate, and a nitrogen base. There are five types of nitrogen bases: adenine, thymine, guanine, cytosine, and uracil. Uracil is only found in the nucleic acid RNA, and thymine is only found in DNA. Nucleotides can be classified as purines or pyrimidines based on their single or double ring structure. Nucleotides are also found in other important biological molecules including adenosine triphosphate (ATP), nicotinamide adenine dinucleotide (NAD^+), and flavin adenine dinucleotide (FAD).

Essential Knowledge covered
3.A.1: DNA, and in some cases RNA, is the primary source of heritable information.
4.A.1: The subcomponents of biological molecules and their sequence determine the properties of that molecule.

Review It

List two major chemical differences between RNA and DNA.

Fill out the following chart on with information about the nitrogen bases found in nucleotides.

	Purine	**Pyrimidines**
Structure		
Number of different types		
Names of representative bases		

Fill out the following functions for the different types of RNA molecules.

RNA molecule	Function
mRNA	
rRNA	
tRNA	

Use It

Draw a nucleotide. Be sure to include the five-carbon sugar, phosphate, and nitrogen base.

What is the name of the nucleotide in your drawing?

Is the nucleotide in your drawing found in DNA, RNA or both?

Do purines form bonds with other purines or with pyrimidines? Provide an example of correct base pairing.

Describe the structure of DNA. Include the terms: *double helix*, *sugar phosphate backbone*, *base pairing*, *3′ end* and *5′ end* in your answer.

Recall It

Proteins are molecules that play a role in many functions including: catalyzing reactions, defense, transport, support, motion, regulation, and storage. Proteins are formed from combinations of the 20 different amino acids. Amino acids are grouped into five classes: nonpolar, polar uncharged, polar charged, aromatic, and amino acids with special functions. Amino acids are linked together by peptide bonds into long chains called polypeptides. The amino acid sequence that makes up a protein is known as a protein's primary structure. The next level of protein structure, secondary structure, describes if the protein is folded into an α-helix coil or β pleated sheet. Proteins have 3-D shape, known as their tertiary structure. Quaternary structure describes the final arrangement of proteins with multiple polypeptides. Similar tertiary structures found in otherwise dissimilar proteins are known as motifs. Substructures within the tertiary structure of a protein that have functionality are known as domains. Proteins need to fold properly in order to work. Proteins can only fold properly within a narrow range of conditions. When a protein unfolds, it becomes inactive or denatured.

Essential Knowledge covered
4.A.1: The subcomponents of biological molecules and their sequence determine the properties of that molecule.
4.B.1: Interactions between molecules affect their structure and function.

Review It

Name three environmental conditions which may alter the proper folding of a protein.

Determine whether or not the following statements are true or false **(T/F)** about amino acids and proteins:

Motifs can be used to predict protein function.

All amino acids are nonpolar.

X-rays can be used to determine protein shape.

A protein that works at 100°C will most likely work at 10°C.

An alteration of just one single amino acid may have a drastic effect on protein function.

Identify the bonds important to protein structure:

Description	Bond
Bonds which form between different amino acids	
Bonds which form between groups with opposite charges	
Bonds which form between two cysteine side chains	

Use It

Draw a dipeptide bond between two generalized amino acids. Label the amino end, the carboxyl end, and the R groups.

In what state would you expect to find proteins from microorganisms in a jar of pickles, and why?

If a protein contains only five amino acids, how many different possible amino acid sequences are in that one protein?

3.5 Lipids: Hydrophobic Molecules

Recall It

Lipids contain a high number of nonpolar carbon-hydrogen bonds. Triglycerides are lipids made up of fatty acids linked to glycerol and store energy very efficiently. Like all lipids, triglycerides are insoluble in water. When placed in water, triglycerides will aggregate together into micelles. More complex lipids are called phospholipids and contain fatty acids, glycerol, and a polar phosphate group. Phospholipids form the core of all biological membranes. When placed in water, phospholipids form lipid bilayers, two layers of polar (hydrophilic) phosphate heads pointing out, and nonpolar (hydrophobic) tails pointing in.

Essential Knowledge covered
2.B.1: Cell membranes are selectively permeable due to their structure.
4.A.1: The subcomponents of biological molecules and their sequence determine the properties of that molecule.

Review It

List three different kinds of lipids.

Compare the functions of triglycerides and phospholipids in biological systems.

Triglycerides	Phospholipids

Use It

Why are biological membranes not formed out of triglycerides?

3 Chapter Review

Summarize It

1. The following table shows the activity of an enzyme in solutions with varying pHs.

   a. At what pH is this enzyme most stable?

   b. What can you infer about the structure of the enzyme above and below this value?

   c. If this enzyme was extracted from the human body, what would have been a more likely source—blood or stomach fluids? Why?

Enzyme Activity (%)	pH
43%	5.8
66%	6.4
98%	7.2
39%	8.4

2. If fats yield 9 kilocalories of chemical energy per gram (kcal/g), and carbohydrates yield only 4 kcal/g, how many more kilocalories would you gain from eating 10 grams of peanut butter compared to 10 grams of bread?

3. Sickle cell disease is a heritable blood disorder in which the surface of hemoglobin molecules (proteins which carry oxygen in the bloodstream) are misshapen. As a result, the hemoglobin aggregates into long, curved rods that deform the red blood cell. This mutation is often the result of a single nucleotide. Using the diagram below, describe where the mutation occurs and how it effects the hemoglobin.

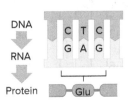

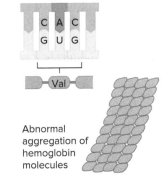

4. The model below shows phospholipids in water. Pose a scientific question that you could test on how the phospholipids would react if placed in a nonpolar solution.

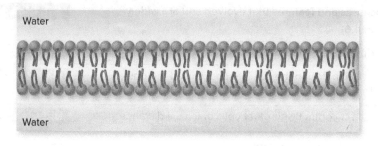

Water

Water

5. Which function is more probable for the following nucleotide sequence?

GCTCCAGGTCA

    a. To store heritable information

    or

    b. To carry genetic information

Justify your answer.

Chapter 4: Cell Structure

Essential Knowledge

1.B.1	Organisms share many conserved core processes and features that evolved and are widely distributed among organisms today. **(4.2, 4.3, 4.6)**	**Big Idea 1**
2.B.1	Cell membranes are selectively permeable due to their structure. **(4.1, 4.2)**	**Big Idea 2**
2.B.3	Eukaryotic cells maintain internal membranes that partition the cell into specialized regions. **(4.2, 4.3)**	**Big Idea 2**
3.D.2	Cells communicate with each other through direct contact with other cells or from a distance via chemical signaling. **(4.8)**	**Big Idea 3**
4.A.2	The structure and function of subcellular components, and their interactions, provide essential cellular processes. **(4.3, 4.4, 4.5)**	**Big Idea 4**

Chapter Overview

Cells are the basic organizational unit for life. All organisms are made up of one or more cells. Cells provide the structural foundation as well as the functional needs for all organisms. Cells vary in complexity but many share similar characteristics.

4.1 Cell Theory

Recall It

The cell theory has three main principals: 1) all organisms are composed of one or more cells, 2) cells are the basic unit of organization for all organisms, and 3) all cells come from other cells. Cells need to remain relatively small so that substances can diffuse in and out of the cell surface. Many different types of cells share similar structures and functions.

Essential Knowledge covered
2.B.1: Cell membranes are selectively permeable due to their structure.

Review It

Name two scientists who played a role in developing the cell theory.

List four variables that may affect rate of diffusion

Identify the following four features all cells have in common:

Feature	Description
	Location of genetic material
	Semifluid matrix in the interior of the cell
	A phospholipid bilayer surrounding the cell
	The site of protein synthesis

Use It

Aegagropila linnaei is a type of filamentous algae that forms spherical balls called marimos under certain conditions. Which marimo would have a greater surface-area-to-volume, a marimo that has a diameter 25 cm or one with a diameter of 5 cm?

Give two reasons why cells remain relatively small.

Copyright © McGraw-Hill Education

4.2 Prokaryotic Cells

Recall It

Prokaryotic cells are relatively simple compared to eukaryotic cells. Prokaryotic cells lack a nucleus, an internal membrane system, and membrane-bounded organelles. In most prokaryotes, DNA is organized into a single circular chromosome. All prokaryotic cells have ribosomes and carry out processes that eukaryotic cells do, such as transcription and translation. In most bacterial prokaryotes, the rigid cell wall surrounding the plasma membrane is composed of peptidoglycan. In archaeal prokaryotes, cell walls are formed from various chemical compounds, including polysaccharides and proteins. Archaea also have different lipids in their membranes compared to bacterial cells.

Essential Knowledge covered
1.B.1: Organisms share many conserved core processes and features that evolved and are widely distributed among organisms today.
2.B.1: Cell membranes are selectively permeable due to their structure.
2.B.3: Eukaryotic cells maintain internal membranes that partition the cell into specialized regions.

Review It

Name the two domains of prokaryotic life.

List three structures found in all prokaryotes.

Use It

Place an *X* next to the cell type where the following structure could be found:

Structure	Human Cell	Bacterial Cell	Archaean Cell
Nucleus			
Ribosome			
Peptidoglycan cell wall			
Polysaccharide cell wall			

Describe how the antibiotic penicillin helps fight a bacterial infection.

Compare and contrast eukaryotic cells to prokaryotic cells.

Recall It

Eukaryotic cells are more complex than prokaryotic cells. Eukaryotic cells remain relatively small which allows for materials to be moved across the membrane with ease. Eukaryotic cells have a membrane-bounded nucleus, an endomembrane system, and many different membrane-bound organelles which carry out specialized functions. The nucleus is the information center of the cell. It is surrounded by two phospholipid bilayers, the outer which is continuous with the endomembrane system. Ribosomes are assembled in the nucleolus and then sent to the cytoplasm, where they translate mRNA to produce polypeptides.

Essential Knowledge covered
1.B.1: Organisms share many conserved core processes and features that evolved and are widely distributed among organisms today.
2.B.3: Eukaryotic cells maintain internal membranes that partition the cell into specialized regions.
4.A.2: The structure and function of subcellular components, and their interactions, provide essential cellular processes.

Review It

Identify the following types of RNA.

RNA	Description
	Carries coding information from DNA
	With protein, makes up ribosomal subunits
	Carries amino acids

List two organelles found in plant cells but not in animal cells.

Provide a definition for the following parts of the nucleus:

Structure	Definition
Nuclear pore	
Nuclear envelope	
Nucleolus	

Use It

Describe two functions of the inner surface of the nuclear envelope, the nuclear lamina.

Using a rough sketch, describe and explain how RNA is moved through the cell to become a protein.

4.4 The Endomembrane System

Recall It

This endomembrane system divides the cell into structural and contractual compartments. The endomembrane system is made up of the following parts: endoplasmic reticulum (ER), the Golgi, and lysosomes. All compartments work together to ensure the cells function efficiently. The ER creates channels and passages within the cytoplasm. There are two types of ER, the rough ER (RER), and the smooth endoplasmic reticulum (SER). The RER is a site of protein synthesis, it also modifies proteins, and manufactures membranes. It is "rough" because it is studded with ribosomes. The SER has many roles, including involvement in carbohydrate and lipid synthesis and detoxification. The function of the Golgi apparatus is to sort and package proteins. It receives vesicles from the ER, modifies and packages macromolecules, and transports them outside of the cell. Lysosomes contain digestive enzymes which break down macromolecules and recycle the components of old organelles.

Essential Knowledge covered
4.A.2: The structure and function of subcellular components, and their interactions, provide essential cellular processes.

Review It

Provide the name or the function of the following organelles:

Structure	Function
Rough ER	
	Collects, packages, and distributes molecules synthesized at one location and used another within the cell
	Aids in the synthesis of molecules such as carbohydrates, steroid hormones, and lipids; also stores intracellular Ca^+
Lysosomes	
	Specialized plant cell organelles which aids in water balance, storage of useful molecules, and waste products

Use It

The diagram to the right has been labeled with letters A-E. Pick two of the five letters and describe the organelle and the process that is occurring at the lettered location.

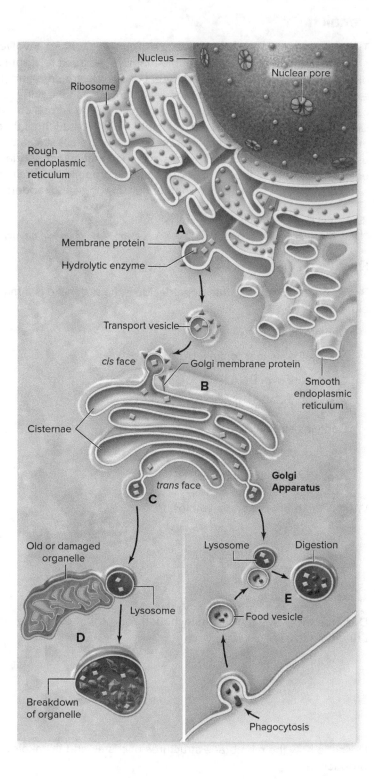

Nucleus

Nuclear pore

Ribosome

Rough endoplasmic reticulum

A

Membrane protein

Hydrolytic enzyme

Transport vesicle

cis face

Golgi membrane protein

Smooth endoplasmic reticulum

B

Cisternae

trans face

C

Golgi Apparatus

Old or damaged organelle

Lysosome

Digestion

E

Lysosome

Food vesicle

D

Breakdown of organelle

Phagocytosis

Recall It

The mitochondria and chloroplasts are organelles that are involved in energy production. They are similar in that they have a double-membrane structure, contain their own DNA, and can divide independently. They are different in overall structure and process. Mitochondria have an extensively folded inner membrane called cristae. On the surface and in the cristae, proteins carry out the metabolism of sugar to produce ATP (the cash currency of our cells' metabolism). Chloroplasts, on the other hand, generate ATP by capturing light energy and oxygen via thylakoid membranes. The thylakoid membranes are arranged in stacks called grana. ATP and CO_2 are used to synthesize glucose. There is evidence that mitochondria and chloroplasts arose by endosymbiosis. The theory of endosymbiosis proposes that mitochondria and chloroplasts were once prokaryotes engulfed by another cell, giving rise to modern-day eukaryotic cells.

Essential Knowledge covered
4.A.2: The structure and function of subcellular components, and their interactions, provide essential cellular processes.

Review It

Based on the following statement, decide if the organelle is a mitochondria **(M)**, chloroplast **(C)**, or either **(E)**:

Contains chlorophyll

Metabolizes sugar to make ATP

Contain its own DNA

Performs photosynthesis

Contains stroma

Has a double membrane

Arose from endosymbiosis

What is the theory of endosymbiosis?

Use It

Describe two ways in which mitochondria and chloroplasts are similar and two ways in which they are different.

Describe how the membrane structures of cristae and thylakoids allow for efficient energy production.

Recall It

The eukaryotic cytoskeleton is a dynamic network of protein fibers. The cytoskeleton supports cell shape, anchors organelles, and moves materials within the cell. It is made up of three different types of fibers: actin filaments, microtubules, and intermediate filaments. Each fiber type has distinct composition and function. Actin filaments are long fibers that are responsible for cell movements, such as crawling, contracting, and pinching during cell replication. Microtubules are formed by rings of tubulin proteins and are organized by barrel-shaped organelles called centromeres. Microtubules are involved in cytoplasm organization, moving materials within the cell, and movement of chromosomes. Finally, intermediate filaments are tough, fibrous proteins which aid in cell stability and structure. The fibers in the cytoskeleton are constantly assembling and dissembling, and interacting with one another, membrane proteins, and vesicles and motor proteins which carry substances through the cell.

Essential Knowledge covered
1.B.1: Organisms share many conserved core processes and features that evolved and are widely distributed among organisms today.

Review It

Identify the fiber of the cytoskeleton based on fiber's composition.

Composition	Fiber
Globular proteins of actin twisted together	
Tetramers of the protein vimentin	
α- and β-tubulin protein subunits arranged side-by-side	

Determine what cytoskeleton fiber is responsible for the following action or trait.

Action/Trait	Fiber
An animal cell pinches into two cells during cellular division.	
A chimpanzee has hard fingernails.	
A sparrow flaps its wing.	
A vesicle is transported along the ER.	
Chromosomes are pulled apart during mitosis.	
A human has hard fingernails.	

Use It

Eukaryotic cells depend on the cytoskeleton for a variety of reasons. Describe three functions of the cytoskeleton.

The cytoskeleton interacts with the endoplasmic reticulum in both animal and fungal cells. Find and describe an example in the text of how the two may interact in either animal or yeast cells.

4.7 Extracellular Structures and Cell Movement

Recall It

Some fibers of the cytoskeleton are responsible for cell movement. Cells crawl as actin filaments polymerize, forcing the edge of a cell forward, and myosin motor proteins on the actin filaments contract, pulling the contents of the cell along. Microtubules and motor proteins called dynein cause undulations to occur in cell projections known as cilia and flagella. Extracellular structures differ in plants and animal cells, but both function to give cells strength, support, and protection. Plants, fungi, and some protists have cell walls made out of polysaccharide cellulose or, in fungi, chitin. Animal cells lack a cell wall but secrete a mixture of proteins, such as collagen, proteoglycans, and fibronectin known as the extracellular matrix (ECM). The ECM can be attached to the plasma membrane and to proteins known as intergrins which link the ECM to the cytoskeleton. Intergrins can influence the overall behavior of the cell through altering gene expression, and a combination of mechanical and chemical signaling pathways.

Review It

Explain the main difference between plant cell walls and fungi cell walls.

The ECM is a mixture of many different type of proteins. Provide the name of one type of protein and its role in the ECM.

Recall It

Cells develop their cellular identity through gene expression. Specific sets of genes are turned on in different types of cells. One key set of genes determines the surface of the cell. Cells that have the same type of surface markers are able to recognize each other as the same type, form connections, send signals, and coordinate their functions. Some cellular surface markers include glycolipids and MHC proteins. Multicellular organisms require cell-to-cell connections or junctions, which give rise to different tissue types. Cell junctions are characterized by their structure and the proteins involved in the junction. Some cell junctions include tight junctions, adhesive junctions, and communicating junctions. Communicating junctions allow the passage of small molecules between cells. In plant cells, communication junctions are known as plasmodesmata, which penetrate the cell wall and connect the cytoplasm of adjoining cells.

Essential Knowledge covered
3.D.2: Cells communicate with each other through direct contact with other cells or from a distance via chemical signaling.

Review It

Provide the function of the following cell-to-cell connections:

Type of connection	Function
Tight junctions	
Surface marker	
Focal adhesion	
Plasmodesmata	
Desmosome	

Explain how cells recognize other cells and the world around them.

Use It

Plant cells are surrounded by rigid structures known as cell walls. How are adjacent plant cells able to communicate to one another or share resources? Use a diagram to help illustrate your answer.

Briefly describe how the vertebrate immune system keeps malignant cells from invading.

4 Chapter Review

Summarize It

1. If the following shapes below represented two different types of cells, which would be better at exchanging molecules (such as nutrients or waste) with the outside environment and why?

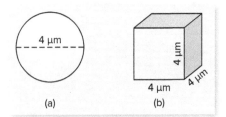

(a)   (b)

2. The diagram below shows (a) the general structure of the virus HIV and (b) how it attaches to the human membrane. Using the diagram below, construct an explanation of how HIV enters a human cell.

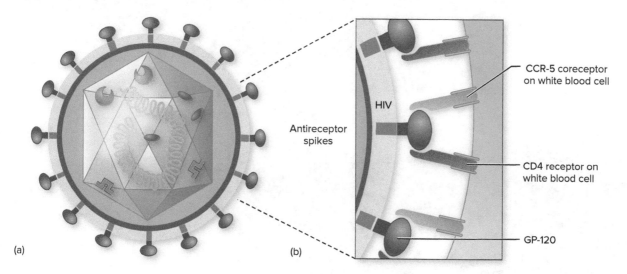

(a)

(b)

Antireceptor spikes

HIV

CCR-5 coreceptor on white blood cell

CD4 receptor on white blood cell

GP-120

3. Why are ribosomes thought of as "universal organelles?"

4. Does the following model represent a prokaryotic or eukaryotic cell?

   Use at least four pieces of evidence to support your answer.

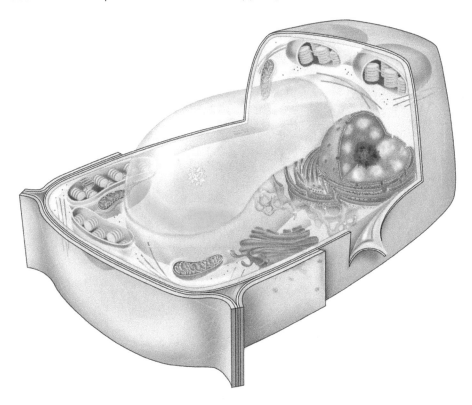

5. If an animal cell needs to remove a damaged peroxisome, predict the steps that it might take to do so.

Chapter 5: Membranes

Essential Knowledge

2.B.1	Cell membranes are selectively permeable due to their structure. (5.1, 5.2, 5.3)	**Big Idea 2**
2.B.2	Growth and dynamic homeostasis are maintained by the constant movement of molecules across membranes. (5.4, 5.5, 5.6)	**Big Idea 2**

Chapter Overview

Cells have phospholipid membranes which allow for protection from the outside world as well as the passage of specific substances and information in and out of the cell. Lipid membranes are also found around the internal structures of some organelles in eukaryotic cells. Substances can pass through membranes through passive transport, active transport, and bulk transport.

5.1   The Structure of Membranes

Recall It

Cell membranes are made up of phospholipid bilayers, transmembrane proteins, interior proteins, and cell-surface markers. The phospholipid bilayer has a hydrophobic inner region and a hydrophilic outer region. Transmembrane proteins float in the lipid bilayer and stiffer internal proteins aid in membrane structural support. Peripheral proteins are associated with the surface of the membrane and include cell-surface markers such as glycoproteins and glycolipids. These four components are represented in the fluid-mosaic model.

Essential Knowledge covered
2.B.1: Cell membranes are selectively permeable due to their structure.

Review It

List the four components found in the fluid mosaic model.

Classify the following as either a peripheral **(P)** or integral protein **(I)**:

    Glycoproteins

    Glycolipids

    Channel proteins

    Carrier proteins

Which direction does the hydrophilic portion of the lipid membrane face?

Use It

Create a diagram of the fluid mosaic model. Be sure to label the membrane's hydrophilic and hydrophobic regions, two types of peripheral membranes, and at least one integral protein.

Describe what might happen to a cell if it lost all proteins in and on its membrane.

5.2 Phospholipids: The Membrane's Foundation

Recall It

Cell membranes can contain lipids including glycerol phospholipids, sphingolipids, and sterols such as cholesterol. Phospholipids have a polar phosphate head with hydrophobic fatty acid tails. In water, phospholipids spontaneously form fluid bilayers. The non-polar interior of a lipid bilayer impedes the passage of water and water-soluble substances. The degree of fluidity of the membrane changes with the composition of the membrane and environmental conditions. The composition can also affect membrane structure.

Essential Knowledge covered
2.B.1: Cell membranes are selectively permeable due to their structure.

Review It

A portion of a cell is shown in the illustration to the right. The cell has been exposed to an aqueous environment.

Correctly label the following parts of the phospholipid membrane:

- Nonpolar interior
- Hydrophilic heads
- Hydrophobic tails

Cytosol

Determine whether or not the following statements are true or false **(T/F)** about cell membranes:

Cell membranes are completely permeable.

All cell membranes are composed of only one type of lipid: the phospholipid.

Cell membranes have a nonpolar interior.

Phospholipids spontaneously form bilayers in water.

Increasing the temperature can make a membrane more fluid.

Use It

Cells are said to be "selectively permeable." In your own words, describe what this means.

How does the structure of the cell membrane allow a cell to be selectively permeable? Use a drawing to help answer the question.

5.3 Proteins: Multifunctional Components

Recall It

Proteins in the cell membrane allow for a cell to interact with its environment and other cells in numerous ways. Transporter proteins allow only certain solutes to leave or enter the cell. Enzymes found on the cell surface aid in chemical reactions. Cell-surface receptors field incoming messages. A cell can be recognized by other cells through cell-surface identity markers. Cell-to-cell adhesion proteins allow for physical connections to be made to other cells. Surface proteins that interact with other cells also create attachments to the cytoskeleton. Transmembrane proteins are anchored in the membrane through regions that span the lipid bilayer called transmembrane domains.

Essential Knowledge covered
2.B.1: Cell membranes are selectively permeable due to their structure.

Review It

Identify the following proteins given their functions:

Protein Class	Function
	Allows another cell to identify its type
	Sticks to another cell
	Allows Na^+ to enter the cell
	Accepts a message from the environment
	Speeds up a reaction occurring on the cell surface

Use It

As described in the last section, the structure of the phospholipid bilayer makes it difficult for water-soluble substances to freely pass through the cell membrane. However, cells need many substances to live. Using the information you learned in this section, describe how a cell may selectively let a water soluble substance in.

5.4 Passive Transport Across Membranes

Recall It

Passive transport occurs when substances move in and out the cell without the cell having to expend energy. One type of passive transport is simple diffusion. Some small non-polar molecules can move across the plasma membrane. These particles move in or out of a cell depending on their concentration on either side of the membrane. Substances that are larger and cannot easily cross the plasma membrane can enter the cell through a process called facilitated diffusion. For facilitated diffusion to occur, a concentration gradient of the substance must still exist, but substances cross the membrane with the help of specific channel or carrier proteins. Water can move across membranes in a process called osmosis and flows through specialized channels for water called aquaporins. Cells in an isotonic solution are in osmotic balance; cells in a hypotonic solution will gain water; and cells in a hypertonic solution will lose water.

Essential Knowledge covered
2.B.2: Growth and dynamic homeostasis are maintained by the constant movement of molecules across membranes.

Review It

Name three types of passive transport.

Identify the following membrane proteins based on their functions.

Function	Protein Type
Bind to a molecule and assist in its diffusion into a cell	
Facilitates the flow of water into a cell	
Provides a hydrophilic chute for polar molecules to pass through into the cell	

Determine if the cell will gain water **(+)**, lose water **(−)**, or stay the same **(Ø)**.

A cell is placed in a hypotonic solution.

A cell is placed in an isotonic solution.

A cell is placed in a hypertonic solution.

Use It

Look at the diagram to the right. Which direction will the ions move through this channel protein and why?

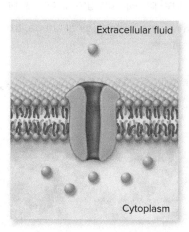

Extracellular fluid

Cytoplasm

If a cell has a pressure potential (Ψ_p) equal to 3 bars and a solute potential (Ψ_s) equal to −0.6 bars, what is the water potential of the cell?

If the cell from the problem above is dropped into a solution that has a water potential of 2 bars, will the free water flow in or out of the cell?

Recall It

Cells can move substances against their concentration gradients but require energy to do so. This process is called active transport. Like facilitated diffusion, active transport requires highly specialized carrier proteins to move selective materials into or out of the cell. Unlike facilitated diffusion, an energy source, usually ATP, is either directly or indirectly involved in the operations. The sodium-potassium pump is an example of a carrier protein that uses ATP directly to operate. In coupled transport, ATP is used indirectly as molecules are moved against their concentration gradient by using the energy stored in a gradient of a different molecule.

Essential Knowledge covered
2.B.2: Growth and dynamic homeostasis are maintained by the constant movement of molecules across membranes.

Review It

Listed below are the six step involved in the Na^+/K^+ pump but some important molecules are missing! Fill in the blanks with Na^+, K^+, or ATP.

Step 1. Three _____ bind to the carrier protein from the intracellular side, causing the protein to change conformation.

Step 2. The protein binds _____, cleaves this into ADP, and is phosphorylated in the process.

Step 3. A second conformational change occurs as the protein is phosphorylated, and the protein translocates the three _____ across the membrane. The three bound _____ break away from the protein and enter the extracellular matrix.

Step 4. The newly conformed protein now has a high affinity for two _____, which bind as soon as the protein is free.

Step 5. The binding of the two _____ cause yet another conformational change, resulting in hydrolysis of the bound phosphate group on the protein.

Step 6. The protein reverts to its original shape, translocating and plopping the two _____ into the interior of the cell.

Name an important molecule transporter which uses the Na^+ gradient produced by the Na^+/K^+ pump as a source of energy.

Use It

If a carrier protein involved in the Na^+/K^+ is able to transport 300 Na^+ outside a cell per second, how many cycles occur per second?

Based on your answer from the question above, how many K$^+$ enter the cell during that second?

Describe the difference between coupled transport and countertransport.

5.6 Bulk Transport by Endocytosis and Exocytosis

Recall It

Large, polar molecules enter and exit cells through bulk transport. Bulk materials enter cells through endocytosis. During endocytosis, substances are enveloped by the plasma membrane. There are three major types of endocytosis: (1) phagocytosis, (2) pinocytosis, and (3) receptor-mediated endocytosis. Cells take in fragments of organic materials through phagocytosis and liquids through pinocytosis. During receptor-mediated endocytosis, endocytosis is initiated by the protein clathrin when target molecules on incoming substances bind to receptors in clathrin-coated pits on the plasma membrane. Exocytosis is the process in which materials leave the cell. Materials are discharged from vesicles through the plasma membrane.

Essential Knowledge covered
2.B.2: Growth and dynamic homeostasis are maintained by the constant movement of molecules across membranes.

Review It

Identify the way in which the following substance may enter or exit the cell.

Substance	Process
Extracellular fluid	
Cellular waste products	
A molecule that binds to a receptor in a clathrin coated pit	
Bacterial cells	

List three substances that may exit an animal cell through exocytosis.

Use It

Choose one of the following types of endocytosis to illustrate: pinocytosis, phagocytosis, or receptor-mediated endocytosis. Label the plasma membrane, the extracellular environment, and the cytoplasm. Describe what type of material is entering the cell and any other pertinent details.

Low-density lipoprotein (LDL) is a molecule that is taken up by receptor-mediated endocytosis. LDL brings cholesterol into the cell so that it can be incorporated into the plasma membrane. What would happen to an individual if they had a mutated form of LDL that didn't fasten to their receptors in the clathrin-coated pits?

5 **Chapter Review**

Summarize It

1. Take a careful look at the following diagram, then pose a scientific question concerning the selective permeability of the cell membrane based on the carrier protein and molecules in the diagram.

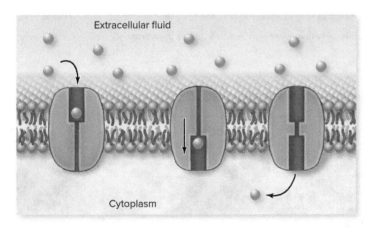

2. Growth factors are substances which can initiate signaling pathways that result in the growth and differentiation of numerous different cell types. Based on the information provided to you in this chapter, create a model that shows how a growth factor (GF) produced in Cell A can tell Cell B to differentiate into another cell type.

3. In intestinal epithelial cells, the active glucose transporter uses the Na^+ gradient produced by the Na^+/K^+ pump to power the movement of glucose into the cells, against its concentration gradient. Given this information, choose one of the following two situations and describe the movement of glucose into the intestinal epithelial cells.

    a. The concentration of glucose outside the epithelial cells is less than the concentration outside of the cells, and the Na^+ concentration outside of the cell is greater than the Na^+ concentration inside of the cell.

    b. The concentration of glucose outside of the epithelial cells is less than the concentration outside of the cells, and the concentration of Na^+ inside the epithelial cells is also less than the concentration of Na^+ outside of the epithelial cells.

Chapter 6: Energy and Metabolism

Essential Knowledge

1.B.1	Organisms share many conserved core processes and features that evolved and are widely distributed among organisms today. **(6.5)**	**Big Idea 1**
2.A.1	All living systems require constant input of free energy. **(6.1, 6.2, 6.3)**	**Big Idea 2**
4.B.1	Interactions between molecules affect their structure and function. **(6.4)**	**Big Idea 4**

Chapter Overview

The laws of thermodynamics govern the flow of energy through living systems. Energy enters the biological world as sunlight and is transformed through chemical and physical means to fuel living systems. At the molecular level, ATP is the main "currency" all cells use for chemical transactions, and these transactions can be regulated by enzymes.

6.1   The Flow of Energy in Living Systems

Recall It

Thermodynamics provide us the laws which govern energy and changes in energy. Energy can come in two forms: kinetic energy and potential energy. Kinetic energy is the energy of motion, and potential energy is stored energy. Energy is commonly measured as heat in biology. Energy mainly enters the biological world through sunlight, where it is converted through photosynthesis into sugars. The energy stored in the bonds of these sugars is used to make new bonds through oxidation-reduction reactions.

Essential Knowledge covered
2.A.1: All living systems require constant input of free energy.

Review It

Determine if the following example is kinetic **(K)** or potential **(P)** energy:

    A snowball being held in a mitten

    A snowball flying through the air

    Water rushing down a hill

    Covalent bonds in sugar molecules

    A crawling baby

Molecules A and B are in a redox reaction. Identify which statement below describes oxidation and which describes reduction.

Molecule A loses an electron during the chemical reaction:
Molecule B gains an electron during the chemical reaction:

Use It

If it takes 98.8 kcals to break one C—H found in an organic molecule, how many kcals does it take to break apart methane, which has a molecular formula of CH_4.

What do you think would happen to life on Earth if the amount of sunlight reaching Earth was reduced by 75%? Why?

6.2 The Laws of Thermodynamics and Free Energy

Recall It

There are two laws of thermodynamics discussed in this chapter. The first law states that energy cannot be created or destroyed; it can only change from one form to another. The second law states that entropy, the disorder in the universe, is continuously increasing. To increase order, energy needs to be expended. The amount of energy available to do work is called free energy (G). If the change in free energy (ΔG) is negative, reactions are spontaneous or exergonic. If ΔG is positive, the reaction is not spontaneous and is called endergonic. Exergonic reactions require activation energy and can be sped up by catalysts whereas endergonic reactions release energy to their surroundings.

Essential Knowledge covered
2.A.1: All living systems require constant input of free energy.

Review It

Determine if the following change in free energy has created
an endergonic **(EN)** or exergonic reaction **(EX)**:

$\Delta G < 0$

$+\Delta G$

$\Delta G > 0$

$-\Delta G$

Define the following terms associated with chemical reactions.

Term	Definition
Catalysis	
Activation energy	
Free energy	

Use It

Underline the following statements that are true about catalysts (more than one statement may be true).

A. Catalysts do not alter the free-energy change produced by a reaction.

B. Catalysts lower the activation energy needed to initiated a reaction.

C. Catalysts violate the basic laws of thermodynamics.

D. Catalysts do not alter the amount of reactant converted into product.

Explain how an embryo can grow larger and more complex as they develop without violating the second law of thermodynamics.

Catalysts are important in driving the chemical reactions occurring in our bodies. Using the empty axes below, sketch the difference in the change of the free energy (G) used over the course of a reaction for both an unanalyzed and a catalyzed reaction. Be sure to label all parts of your graph, including the reactant, product, and activation energy of each reactant.

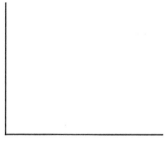

Recall It

Adenosine triphosphate (ATP) is the energy currency of cells. Energy is stored in the bonds of ATP's triphosphate groups. The negative charges of phosphate groups strongly repel one another. When these bonds are broken through hydrolysis, creating ADP, a tremendous amount of energy is released. This energy is used to fuel almost all cellular energy-requiring processes. Cells are constantly building ATP up and breaking ATP down to run endergonic reactions.

Essential Knowledge covered
2.A.1: All living systems require constant input of free energy.

Review It

List the three subcomponents of the nucleotide ATP and circle the one which contains the high energy bonds.

Identify if the following reactions are endergonic or exergonic.

$ADP + P_i \rightarrow ATP$	$ATP \rightarrow ADP + P_i$	a muscle contract

Do cells retain large stockpile ATP?

Use It

Which of the following hypothetical biological reactions would become spontaneous when coupled with ATP hydrolysis? Support your answer mathematically. Recall that ATP hydrolysis has a ΔG of -7.4 kcal/mol.

A. $A + B \rightarrow C$; $\Delta G_C = +2.0$ kcal/mol

B. $C + D \rightarrow E$; $\Delta G_E = +8.3$ kcal/mol

C. $E + F \rightarrow G$; $\Delta G_G = -3.0$ kcal/mol

Provide at least two specific examples of biological process for which ATP provides energy.

6.4 Enzymes: Biological Catalysts

Recall It

Catalysts regulate chemical reactions. In biological systems, most catalysts are enzymes, although some RNA molecules have catalytic activities. Catalysts lower the activation energy needed for a reaction to occur by bringing two substrates together in the correct orientation or by stressing particular chemical bonds. There are many different types of enzymes, each which catalyze specific chemical reactions. Enzymes are sensitive to environmental factors, including pH and temperature, as well as the presence of regulatory molecules. Regulatory molecules are substances that decrease or increase enzyme activity. Some enzymes may also require cofactors or coenzymes in order to function.

Essential Knowledge covered
4.B.1: Interactions between molecules affect their structure and function.

Review It

List two ways an inhibitor can decrease enzyme activity.

Determine whether or not the following factors affect the rate of an enzymatic reaction. Place a + if the rate of reaction increases, a − if the rate decreases, or an = sign if the rate remains constant.

Factor	Rate of Reaction
An inhibitor is present	
The concentration of substrate decreases	
The temperature has increased	
Optimal pH has been obtained	

What is the optimal working temperature for most enzymes found in the human body?

Use It

Will the following substrate fit into the following enzyme? If not, describe any changes that may need to occur in order for the substrate to fit.

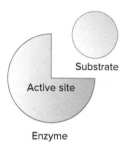

Substrate

Active site

Enzyme

Tryptophan is an amino acid found in the human body. It is needed to produce the neurotransmitter serotonin, a chemical found in our brains associated with happiness. Lower levels of serotonin have been linked to cases of depression. It has also been found that low levels of the small organic molecule Vitamin B_6 are associated with symptoms of depression. Assuming the two are linked, predict the role of Vitamin B_6 in serotonin's synthesis.

Recall It

The total of all chemical reactions carried out in an organism is called metabolism. Metabolism is made up of two parts: anabolism and catabolism. Anabolism is the energy expended to build molecules, and catabolism is the energy harvested as molecules are broken down. Metabolism is organized through biochemical pathways. Biochemical pathways are series of chemical reactions. Some biochemical pathways are regulated by feedback inhibition, in which the end-product of the pathway stops the pathway from initiating again.

Essential Knowledge covered
1.B.1: Organisms share many conserved core processes and features that evolved and are widely distributed among organisms today.

Review It

Are the following examples of anabolism or catabolism?

Glucose is broken down during cellular respiration.

Plants produce sugars through photosynthesis.

Determine if the following statements regarding biochemical pathways are true or false **(T/F)**:

Biochemical pathways organize biological reactions.

The steps in a biochemical pathway probably evolved all at once.

The first biochemical pathways most likely scavenged energy-rich molecules from the environment.

Regulation of a biochemical pathway often depends on a feedback mechanism of some kind.

A biochemical pathway with feedback inhibition will continue indefinitely.

Use It

Use the biochemical pathway below to answer the following questions. This simple biochemical pathway, contains two enzymes (E) that are needed in order to complete a reaction:

$$A + B \xrightarrow{E_1} C \xrightarrow{E_2} D + F \rightarrow G$$

If E_1 became denatured, what would happen to the pathway?

Which part of the reaction most likely evolved first?

If product G bound to an allosteric site on E_1, describe what would happen to the pathway. What is the name for that kind of mechanism?

Summarize It

1. Phytoplankton are single-celled photosynthesizers that make up the base of the food web for much of the ocean. In 2010, a major oil spill occurred in the Gulf of Mexico. The company responsible for the spill then employed a common method to clean up the oil spill, adding a dispersant to the ocean to help break the oil up into tiny droplets. An independent study later investigated the impact of these substances on phytoplankton in a lab and found that populations of phytoplankton decrease greatly when introduced to the combination of oil and dispersants.

   Provided this information, predict how might the oil spill effect the free energy available in the Gulf of Mexico.

2. The following reaction rates were observed for a reaction under different conditions:

Condition	Reaction Rate
Reaction without enzymes	3.0×10^{-4} mol L^{-1} sec^{-1}
Reaction with enzyme A	0.2×10^{-4} mol L^{-1} sec^{-1}
Reaction with enzyme B	1.2×10^{-4} mol L^{-1} sec^{-1}

Which reaction proceeded the fastest, and what can you determine concerning the different enzymes used in this reaction?

3. Not all enzymes are proteins. In the early 1980s, scientists began to realize that the informational molecule RNA also can function as a catalyst. What does this information suggest concerning the evolution of protein and RNA?

4. During childbirth, a woman's brain releases the hormone oxytocin. This molecule stimulates a contraction in the uterus. If this biochemical pathway is not inhibited by a feedback system, describe what might happen next.

Chapter 7: How Cells Harvest Energy

Essential Knowledge

1.B.1	Organisms share many conserved core processes and features that evolved and are widely distributed among organisms today. **(7.1)**	Big Idea 1
1.D.2	Scientific evidence from many different disciplines supports models of the origin of life. **(7.10)**	Big Idea 1
2.A.1	All living systems require constant input of free energy. **(7.1, 7.2, 7.4)**	Big Idea 2
2.A.2	Organisms capture and store free energy for use in biological processes. **(7.2, 7.3, 7.4, 7.5, 7.8, 7.9)**	Big Idea 2
4.A.2	The structure and function of subcellular components, and their interactions, provide essential cellular processes. **(7.5)**	Big Idea 4
4.B.1	Interactions between molecules affect their structure and function. **(7.7)**	Big Idea 4

Chapter Overview

Cells harvest energy through cellular respiration. Cellular respiration is a universal process, in which chemical bonds of organic molecules, such as glucose, are broken down through a series of redox reactions into carbon dioxide, water, and ATP. Cellular respiration can be aerobic or anaerobic, and can be described through a series of metabolic reactions.

7.1 | Overview of Respiration

Recall It

The role of respiration is to provide energy to the cell. Cellular respiration occurs aerobically (with oxygen) and anaerobically (without oxygen). Electron carriers play a critical role in cellular respiration, carrying energy throughout the system. The electron transport chain in the mitochondria of eukaryotic cells is used to move electrons in order to capture energy efficiently. The ultimate goal of cellular respiration is synthesis of ATP. ATP is then used to power most of the cell's activities. Cells can make ATP through two different mechanisms: (1) Substrate-level phosphorylation, in which phosphate is directly transferred to ADP, and (2) oxidative phosphorylation which generates ATP via the enzyme ATP synthase.

Essential Knowledge covered
1.B.1: Organisms share many conserved core processes and features that evolved and are widely distributed among organisms today.
2.A.1: All living systems require constant input of free energy.
2.A.2: Organisms capture and store free energy for use in biological processes.

Review It

Write the chemical equation for cellular respiration. Identify which molecules are oxidized and which are reduced.

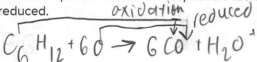

List three types of electron carriers and how they transport electrons.

- soluble carriers move electrons from one molecule to another
- membrane-bound carriers, form a redox chain
- carriers that move within the membrane

Name two charged particles NAD$^+$ can carry.

nicotinamide & monophosphate

Use It

Why does respiration occur in steps and not all at once?

if all the steps occured at once the cells would recover very little of the energy in a useful form

How does ATP synthase catalyze the following reaction?

$$ADP + P_i \rightarrow ATP$$

$$ADP + P_i \xrightarrow{2e^-} ATP$$

by ATP using energy from a proton gradient.

7.2 Glycolysis: Splitting Glucose

Recall It

Glycolysis is the stage in cellular respiration where glucose is converted into two pyruvate molecules, producing two molecules of ATP and NADH in the process. Glycolysis occurs in a series of reactions. The first five reactions require ATP to convert a molecule of glucose into two molecules of glyceraldehyde 3-phosphate (G3P). The second set of reactions covert G3P into pyruvate through the oxidation of G3P, reducing NAD$^+$ to NADH. In the presence of oxygen, NAD$^+$ is regenerated in the electron transport chain. A fermentation reaction is required in the absence of oxygen to regenerate NAD$^+$ through the reduction of pyruvate. Pyruvate is used in aerobic respiration to produce acetyl groups, which are needed to produce ATP in the Krebs cycle.

Essential Knowledge covered
2.A.1: All living systems require constant input of free energy.
2.A.2: Organisms capture and store free energy for use in biological processes.

Review It

Determine the number of molecules produced during the process of glycolysis:

ATP   2

Pyruvate   2

NADH   2

Circle the reaction that occurs in the absence of oxygen.

Aerobic respiration   Fermentation

Describe the location of where glycolysis occurs in the cell.

the glycolysis occurs in the cytosol

Use It

Substrate-level ATP synthesis occurs in the later steps of glycolysis. Draw a picture of how an enzyme might transfer phosphate to ADP to form ATP. Be sure to label the enzyme and molecules.

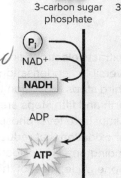 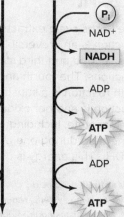

Glycolysis occurs in a series of reactions, as shown on the right. Take a look now at the series of different reactions. Consider the inputs and outputs of the different reactions. Which reaction do you think evolved first based on your knowledge of biochemical pathways?

the second half of glycolysis the ATP breakdown of G3B evolved first. Many biochemical pathways evolved backwards

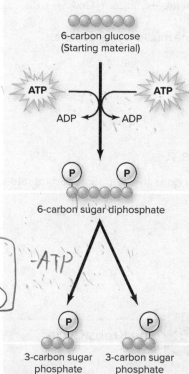

6-carbon glucose (Starting material)

ATP   ATP

ADP   ADP

P   P

6-carbon sugar diphosphate

−ATP

P   P

3-carbon sugar phosphate   3-carbon sugar phosphate

Pi   Pi

NAD+   NAD+

NADH   NADH

ADP   ADP

ATP   ATP

ADP   ADP

ATP   ATP

3-carbon pyruvate   3-carbon pyruvate

Recall It

The pyruvate produced by glycolysis can be further oxidized in the presence of oxygen. Pyruvate is oxidized in the mitochondria, where it yields one molecule of CO_2, one NADH, and one acetyl-CoA. Acetyl-CoA feeds acetyl groups into the Krebs cycle.

Essential Knowledge covered
2.A.2: Organisms capture and store free energy for use in biological processes.

Review It

Fill in the blanks to complete the oxidation of pyruvate.

Pyruvate + _NAD⁺_ + CoA → _acetyl_ + NADH + _CO₂_ + H⁺

Define multienzyme complex. *organizes a series of enzymatic steps so that the chemical intermediates do not diffuse away or undergo other reactions*

Use It

How does the oxidation of pyruvate link glycolysis and the Krebs cycle? *Pyruvate is oxidized in the mitochondria to produce acetyl-Coa and CO_2. Acyetyl-COA is the molecule that links glycolysis to the reaction of the krebs cycle*

Recall It

The Krebs cycle extracts electrons and synthesizes one ATP in a series of nine reactions. The first reaction is an irreversible condensation that produces citrate; it is inhibited when ATP is plentiful. The second and third steps reposition the hydroxyl group on the citrate to allow for subsequent reactions. The fourth and fifth steps are oxidations, both of which reduce NAD^+ to NADH. The sixth reaction is a substrate-level phosphorylation producing GTP, and from that ATP. The seventh reaction is another oxidation that reduces FAD to $FADH_2$. Reactions eight and nine regenerate oxaloacetate, including one final oxidation that reduces NAD^+ to NADH. While only one ATP is generated during one cycle, most of the energy is retained in form of the electrons in NADH and $FADH_2$. This energy is used to generate a proton gradient to drive ATP synthesis.

Essential Knowledge covered
2.A.1: All living systems require constant input of free energy.
2.A.2: Organisms capture and store free energy for use in biological processes.

Review It

Identify the parts in the diagram of the Krebs Cycle with the following symbols:

An enzyme	●
ATP synthesis	♥
CO_2	▼
Electron carriers	■

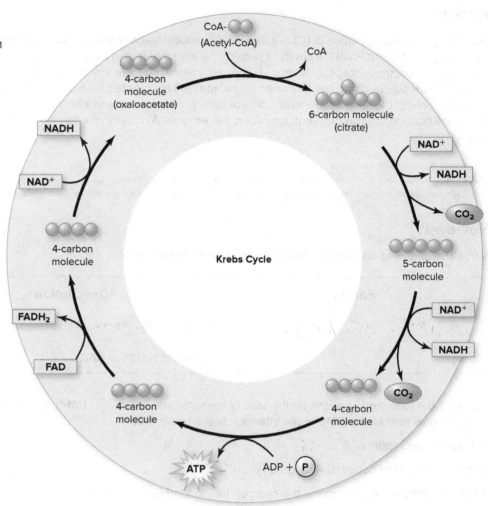

Use It

Through cellular respiration, glucose is broken down and turned into energy. Identify the energy produced in the Krebs cycle.

the energy produced in cellular respiration is ATP, there is also some free energy and heat is produced through this process (entropy), ATP is main energy source.

Underline any FALSE statements regarding the Krebs cycle.

A. <u>Carbon dioxide is consumed during the Krebs Cycle.</u>

B. <u>The sixth reaction of the Krebs Cycle is optional.</u>

C. There are many enzymes that make the Krebs Cycle possible.

D. The Krebs cycle reactions take place in the mitochondrial matrix.

How and where is ATP formed during the Krebs Cycle?

sixth reaction of krebs cycle. takes ADP & Pi converting it to ATP.

Recall It

The electron transport chain (ETC) is a series of membrane-associated proteins. Electrons carried by NADH and $FADH_2$ from the Krebs Cycle are transferred along these proteins toward the terminal electron receptor, oxygen. The energy released from the electron transfer allows for protons to be pumped into the intermembrane space. This creates an electrochemical gradient. Here, the process known as chemiosmosis occurs; protons move back across the membrane, down their concentration gradient, powering the enzyme ATP synthase to phosphorylate ADP into ATP.

Essential Knowledge covered
2.A.2: Organisms capture and store free energy for use in biological processes.
4.A.2: The structure and function of subcellular components, and their interactions, provide essential cellular processes.

Review It

Name the following enzymes found in the electron transport chain.

Enzyme	Description
ATP synthase	Produces ATP from ADP and P_i
Dehydrogenase	Oxidizes NADH to NAD^+

Identify if the following happens on the side of the mitochondrial matrix **(MM)**, the inner mitochondrial membrane **(IM)**, or the intermembrane space **(IS)**:

ATP synthase creates ATP. (IM)

Protons form a high concentration gradient. (IM)

$FADH_2$ contributes an electron to the electron transport chain (MM)

NADH contributes an electron to the electron transport chain. (IM)

Water is formed. (IM)

Use It

There are many electron carries in the electron transport chain. What is the terminal acceptor in aerobic systems?

oxygen

Can ATP synthase work without a proton gradient? Why or why not?

no because the proton gradient produces the energy that drives the rotation of ATP synthase generator.

Recall It

The theoretical yield of ATP harvested from glucose by aerobic respiration is 32 molecules of ATP. This number is reduced to 30 in eukaryotes due to the cost of NADH transport into the mitochondria.

Review It

The formation of 30 to 32 ATP molecules are theoretically possible when glucose is broken down completely. How many ATP are produced during glycolysis?

2 ATP

If a bacterial completely broke down 3 molecules of glucose, what would be the prokaryote's the theoretical yield of ATP?

3 glucose × 2 ATP = 6 ATP

If a human cell broke these three glucose molecules, would it produce more or less ATP than the bacterial cell?

less, glycolysis must be transported

7.7 Regulation of Aerobic Respiration

Recall It

Glucose catabolism is controlled by the concentration of ATP molecules. When levels of ATP are high, key reactions of cellular respiration are inhibited. In this way, ATP is an allosteric inhibitor. When ATP levels are low, ADP activates enzymes in the pathway to being producing more ATP.

Essential Knowledge covered
4.B.1: Interactions between molecules affect their structure and function.

Review It

List the two key points along the biochemical pathway of cellular respiration where the reaction can be inhibited

Glycolysis & ATP

Use It

How are levels of ATP in the biochemical pathway for in glucose metabolism an example of feedback inhibition? *it closes pathways for travel*

7.8 Oxidation Without O$_2$

Recall It

Some organisms live in areas that lack oxygen. These organisms can still respire anaerobically, using inorganic molecules as final electron acceptors in place of oxygen in the electron transport chain. Other organisms use fermentation which uses organic compounds as electron acceptors. Fermentation is the regeneration of NAD$^+$ by oxidation of NADH and reduction of an organic molecule. Unlike in animals where pyruvate is reduced directly to lactate and stored in the muscles, in yeast pyruvate is decarboxylated, then reduced to ethanol.

Essential Knowledge covered
2.A.2: Organisms capture and store free energy for use in biological processes.

Review It

Provide the final electron acceptor in respiration for following organisms:

Organism	Electron acceptor
Yeast	NAD+
Methanogens	NAD+
Sulfur bacteria	NADH

Name two types of fermentation that eukaryotic cells are capable of performing.

lactic acid, achlohol

Use It

Lactic acid fermentation and ethanol fermentation both reduce a metabolite of glucose, oxidizing NADH back to NAD$^+$. How are the end products of these two reactions similar and how are they different?

Recall It

Proteins and fats are important sources of energy. The catabolism of proteins breaks down amino acids and then removes amino groups. The catabolism of fatty acids occurs through conversion of fatty acids into acetyl groups through successive oxidations. These acetyl groups are then fed into the Krebs cycle to be oxidized and generate NADH for electron transport and ATP production.

Essential Knowledge covered
2.A.2: Organisms capture and store free energy for use in biological processes.

Review It

Name two processes which receive energy from protein catabolism.

List two key intermediates which connect the oxidation of food molecules to metabolism.

Use It

Describe the macromolecules you could metabolize from eating a slice of pizza with peppers, onions, and sausage.

Recall It

The stages of metabolism evolved over time. The most primitive life forms probably obtained carbon containing molecules that were abiotically produced, then began storing this energy in the bonds of ATP. Glycolysis most likely followed shortly after. The third major event in the evolution of metabolism was anoxygenic photosynthesis, followed by oxygen-forming photosynthesis. This paved the way for nitrogen fixation and aerobic respiration.

Essential Knowledge covered
1.D.2: Scientific evidence from many different disciplines support models of the origin of life.

Review It

Place the following evolutionary events of metabolism in chronological order **(1–6)**:

Anoxygenic photosynthesis

Storage of energy in ATP

Nitrogen fixation

Oxygen-forming photosynthesis

Glycolysis

Aerobic respiration

Use It

Provide two pieces of evidence that support the idea that aerobic respiration evolved after photosynthesis during evolution of metabolism.

7 Chapter Review

Summarize It

1. How does glycolysis support the concept of common ancestry for all organisms?

2. Earth's atmosphere is now 20.9% oxygen. Did Earth's atmosphere always contain this much oxygen? What data is available to us to answer this question?

3. A sample of bacteria was taken from a sulfur-containing hot spring and spread on a petri dish rich with glucose to grow in the lab. The bacteria died. Why was this bacteria unable to survive?

4. A scientist was studying an organism that had inadequate ATP output during metabolism and isolated the issue to the electron transport chain. What question might the scientist pose to determine where in the electron transport chain was defective?

5. Cyanide is a poisonous substance because it can bind to cytochrome, one of the membrane proteins found in the electron transport chain. Why would this be dangerous?

Chapter 8: Photosynthesis

Essential Knowledge

2.A.1	All living systems require constant input of free energy. (8.1, 8.6)	**Big Idea 2**
2.A.2	Organisms capture and store free energy for use in biological processes. (8.5, 8.6)	**Big Idea 2**
4.A.2	The structure and function of subcellular components, and their interactions, provide essential cellular processes. (8.1, 8.3, 8.4)	**Big Idea 4**
4.A.6	Interactions among living systems and with their environment result in the movement of matter and energy. (8.7)	**Big Idea 4**
4.C.1	Variation in molecular units provides cells with a wider range of functions. (8.3)	**Big Idea 4**

Chapter Overview

Photosynthesis is the process used by some organisms to capture energy from sunlight and convert it into chemical energy. Like respiration, photosynthesis can be described in a series of biochemical reactions and cycles. The process of photosynthesis evolved in bacteria, algae, and plants, and allowed for the diversity of life as we know it on Earth, the oxygen we breathe, and even the energy we extract from food molecules.

8.1 Overview of Photosynthesis

Recall It

Photosynthesis is the conversion of light energy into chemical energy. Photosynthesis takes place in special organelles called chloroplasts, which contain membranes called thylakoids. Within the thylakoids, there are photosystems which capture light energy. Photosynthesis combines CO_2 and H_2O to produce glucose and O_2. Photosynthesis has three stages: (1) absorbing light energy, (2) using this energy to synthesize ATP and NADPH, and (3) using the ATP and NADPH to convert CO_2 to organic molecules. The first two stages compose the light-dependent reactions. The third stage take place in light-independent reactions.

Essential Knowledge covered
2.A.1: All living systems require constant input of free energy.
4.A.2: The structure and function of subcellular components, and their interactions, provide essential cellular processes.

Review It

Fill in the blanks to complete the equation that summarizes photosynthesis:

$6CO_2 + 12H_2O + \underline{light} \rightarrow \underline{C_6H_{12}O_6} + 6H_2O + \underline{CO_2}$

Describe the stages of photosynthesis that happen in the following reactions:

Light-Dependent Reactions	Light-Independent Reactions
• capturing energy from sunlight • using energy to make ATP and to reduce the compound NADP+, an electron carrier, to NADPH	• using the ATP and NADPH to power the synthesis of organic molecules from CO_2 in the air

Describe the structures of a chloroplast.

Structure	Description
stroma	Semiliquid substance surrounding the thylakoid membrane
chlorophyll-a	Photosynthetic pigment that captures light energy
thylakoid membrane	The internal membrane of chloroplasts

Use It

In the diagram below, fill in the missing energy sources and reactions that are occurring.

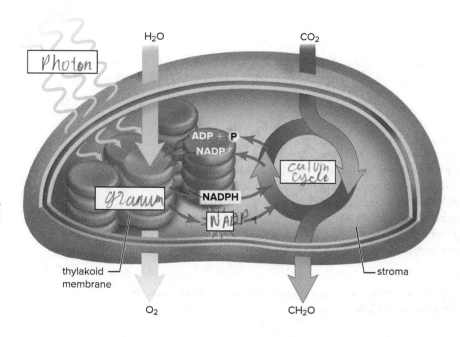

What does the diagram represent? Provide a description of the reactions occurring in your answer.

photosynthesis, photon strikes system breaks water in 1/2 O₂ & 2H. photons go into chloroplast and the thylacoid absorb photons. the energy is used to make ATP and NADPH

8.2 The Discovery of Photosynthetic Processes

Recall It

In the late 1600s, Jan Baptitsa was the first person to scientifically test how plants grow, but he incorrectly deduced the reason for the plant's increase of mass. Well over 100 years later, it was found that plants use sunlight to spilt carbon dioxide into carbon and oxygen and create carbohydrates in the process. The complexity of photosynthesis wasn't fully elucidated until the 1950's after a series of more modern scientific tests were performed, included those with radiolabeled oxygen.

Review It

Robin Hill was a plant biochemist who performed a series of experiments that had extreme important significance for our understanding of photosynthesis. List two important outcomes of Hill's work.

firmly demonstrates plant photosynthesis occur in chloroplast

---

8.3 Pigments

Recall It

Light is a form of energy that exists both as a wave and a particle. Particles of light are known as photons. The energy in light is inversely proportional to the wavelength of the light. Light can remove electrons from some molecules through a process called the photoelectric effect. In photosynthesis, chloroplasts act as photoelectric devices; absorbing light and transferring energy as electrons to a carrier. There are many types of pigments which absorb light. Chlorophyll *a* is the only pigment that can convert light energy into chemical energy. Chlorophyll *b*, carotenoids, and other accessory pigments increase a plant's ability to harvest photons for photosynthesis.

Essential Knowledge covered
4.A.2: The structure and function of subcellular components, and their interactions, provide essential cellular processes.
4.C.1: Variation in molecular units provides cells with a wide range of functions.

Review It

The graph below shows the absorption spectra for the two most common forms of photosynthetic pigment and a group of accessory pigments.

Create a key for the graph.

———— chlorophyll a

- - - - chlorophyll b

——— caratenoids

What color light do chlorophyll pigments absorb?

violet, blue, red light

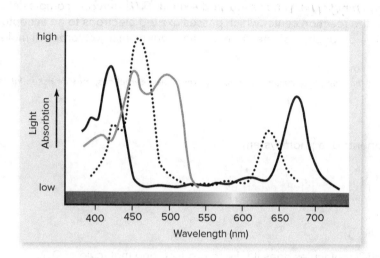

What color light do cartenoid pigments absorb?

blue-green

Use It

Why are the leaves of an oak tree green in the summer and red in the fall?

cool fall temp. causes leaves to cease manufacturing chlorophyll chlorophyll no longer present to reflect green light, and the leaves reflect orange and yellow light that caratenoids and other pigments don't

Plants contain different pigments which absorb light energy at different wavelengths. What allows proteins to absorb specific wavelengths?

phycobiliproteins allow proteins to absorb specific wavelengths

Recall It

In chloroplasts, light is captured by photosystems. Photosystems contain an antenna complex of hundreds of pigment molecules which gather energy from photons and a reaction center which consist of chlorophyll a molecules and associated proteins. The antenna complex channels the collected energy to the reaction center which passes excited electrons to an acceptor outside of the photosystem. The production of one O_2 molecule requires many chlorophyll molecules.

Essential Knowledge covered
4.A.2: The structure and function of subcellular components, and their interactions, provide essential cellular processes.

Review It

List the two components of a photosystem.

antenna complex
carotenoids

How many chlorophyll molecules does it take to produce one molecule of O_2?

2,500

Use It

How does the reaction center move the energy absorbed from photons away from chlorophyll to aid in the conversion of light into chemical energy?

Recall It

The light-dependent reactions can be broken down into four stages: (1) primary photoevent, (2) charge separation, (3) electron transport, and (4) chemiosmosis. In photosynthetic bacteria, this process occurs in a single photosystem. The chloroplasts of plants and algae have two photosystems which play a role in the light-dependent reactions through noncyclic photophosphorylation. Photosystem II and photosystem I are connected by an electron transport chain that pumps protons into the thylakoid space, creating a proton gradient which is used to synthesize ATP. Photosystem I transfers electrons to $NADP^+$ reducing it to NADPH. Photosystem II replaces electrons lost by photosystem I. Electrons lost from photosystem II are replaced by electrons from oxidation of water, which also produces O_2. Photosystem II is primarily found in the grana, while photosystem I and ATP synthase are found in the stroma. Plants can make additional ATP by cyclic photophosphorylation.

Essential Knowledge covered
2.A.2: Organisms capture and store free energy for use in biological processes.

Review It

Describe what happens in each stage of the light-dependent reactions:

Primary photoevent	Charge separation	Electron transport	Chemiosmosis
photon of light captured in pigment	transferred to reaction center, transfers to primary acceptor	shuttled along a series of electron carrier molecules	flow back across membrane through ATP synthase

Identify the photosystems based on their function. Circle the photosystem that most closely resembles the reaction center of purple bacteria.

Photosystem	Function
PS II	Absorbs photons, exciting electrons that are passed to PQ. Site of water oxidation.
PS I	Absorbs photons, exciting electrons that are passed through a carrier to reduce $NADP^+$ to NADPH

Use It

Draw a picture to show how photosystem I and II are connected.

Photosystem I and II are said to work together in "noncyclic photophosphorylation." What does this term mean? Would this process be able to work without water?

means they are used to both produce ATP and NADPH. cant work without water

Provide an explanation for how a proton gradient is formed across the thylakoid membrane during photosynthesis.

protons pumped from stroma to thylakoid compartment.

How is ATP production similar in cellular respiration and photosynthesis? How is it different?

Cell resp. uses glucose and O_2 to make CO_2 and H_2O. Photosynthesis uses photon energy, H_2O, and CO_2 to produce glucose and O_2.

8.6 Carbon Fixation: The Calvin Cycle

Recall It

Calvin cycle reactions convert inorganic carbon into organic molecules. The Calvin cycle uses CO_2, ATP, and NADPH to build simple sugars. The Calvin cycle occurs in three stages: (1) carbon fixation, (2) reduction of the resulting 3-carbon PGA to G3P, generating ATP and NADPH, and (3) the regeneration of RuBP. Six turns of the cycle fix enough carbon to produce two excess G3Ps which can be combined to make one molecule of glucose.

Essential Knowledge covered
2.A.1: All living systems require constant input of free energy.
2.A.2: Organisms capture and store free energy for use in biological processes.

Review It

Fill in the blanks to complete the simple equation that summarizes the Calvin cycle:

$\underline{6 CO_2}$ + 18ATP + $\underline{12 NADPH}$ + water → 2 G3P + 16 P_i + $\underline{18 ADP}$ + $\underline{12 NADP^+}$

How many G3P does it take to make one glucose molecule?

2

How many turns of the Calvin cycle does it take to make one molecule of glucose?

6 turns

Use It

If the Calvin cycle went through 24 turns, how many molecules of glucose could theoretically be formed?

4 glucose

Where does the ATP and NADPH used in the Calvin cycle come from and what is it used for?

come from light used on calvin to reduce
CO_2 into sugar used to convert CO_2 to sugar
from ATP & NADPH

What happens to the output of the Calvin cycle?

Glucose, producing ATP and pyruvate then
transferring to krebs cycle

8.7 Photorespiration

Recall It

Photorespiration occurs when CO_2 concentrations are low inside of a plant leaf, causing the enzyme rubisco to erroneously bind to O_2 instead of CO_2. This commonly occurs in plants in hot, arid environments that close their stomata to conserve water. Alternate methods of photosynthesis have evolved in plant to minimize photorespiration compared to the Calvin cycle alone (C_3 photosynthesis). By adding CO_2 to a 3-carbon molecule, forming oxaloacetate, some plants fix carbon in a process called C_4 photosynthesis. Carbon is fixed in one cell by the C_4 pathway, then CO_2 is released in another cell for the Calvin cycle. The Crassulacean acid metabolism (CAM) pathway divides the light and dark reactions of photosynthesis into night and day. CAM plants use the C_4 pathway during the day when stomata are closed, and the Calvin cycle at night in the same cell. C_4 and CAM pathways also use the enzyme PEP carboxylase.

Essential Knowledge covered
4.A.6: Interactions among living systems and with their environment result in the movement of matter and energy.

Review It

Determine whether or not the following statements are true or false **(T/F)** about the different types of photosynthesis:

C_4 plants partition carbon fixation and the Calvin cycle in different spaces.

Plants with CAM photosynthesis are most likely found in environments where temperatures are below 25°C.

Plants with CAM photosynthesis partition the fixation of CO_2 temporally.

C_3 plants use the enzyme PEP carboxylase to form a 4-carbon molecule

Plants that fix carbon using only the Calvin cycle are called C_3 plants

Provide two examples of both C_4 and CAM plants.

Use It

Using either C_4 or CAM photosynthesis as an example, explain how plants have adapted to survive in hot, arid climates through minimizing the loss of energy through photorespiration. Consider all physical, temporal, and spatial changes of the biochemical pathway of photosynthesis in your answer.

Summarize It

1. A scientist was cataloguing plants in a botanical garden. Given the data sheet below, predict the type of photosynthesis the plant uses.

> Acquisition No.: 583
> Plant name: *Sedum morganianum*
> Origin: Southern Mexico
> Description: Succulent with round, short fleshy leaves
> Temperature required for growth: >25°C
> Watering needs: Infrequent

2. Given the following dataset, would this plant grow better in a room that was constantly full of people or in a room that was only occupied by one person? Explain your answer.

Biomass (g)	CO_2 levels (ppm)
11.1	200
23.4	400
43.5	600

3. A plant physiologist named F.F. Blackman studied the effect of temperature on photosynthesis. He found that photosynthesis could be increased by increasing temperatures, but the rate only increased until about 35°C. Using what you learned about biochemical reactions in Chapter 6, what can you infer about photosynthetic reactions?

4. If the climate surrounding a desert shifted over time so that temperatures lowered from 35°C to 24°C, how might this affect desert plant communities? What would you predict if the temperatures increased from 35°C to 49°C?

5. Chlorophyll *a*, chlorophyll *b*, and carotenoids are pigments that can all be found in a single plant. If chlorophyll *a* is the main photosynthetic pigment, from an evolutionary standpoint, why do plants still have other pigments?

Chapter 9: Cell Communication

Essential Knowledge

3.B.2	A variety of intercellular and intracellular signal transmissions mediate gene expression. **(9.3, 9.4)**	**Big Idea 3**
3.D.1	Cell communication processes share common features that reflect a shared evolutionary history. **(9.1, 9.4)**	**Big Idea 3**
3.D.2	Cells communicate with each other through direct contact with other cells or from a distance via chemical signaling. **(9.1)**	**Big Idea 3**
3.D.3	Signal transduction pathways link signal reception with cellular response. **(9.2, 9.4, 9.5)**	**Big Idea 3**
3.D.4	Changes in signal transduction pathways can alter cellular response. **(9.5)**	**Big Idea 3**

Chapter Overview

Cells communicate through signal transduction. Cells require signal molecules to bind to receptor proteins in order to receive messages. Receptors can be found on cell membranes or inside cells, and binding signals can cause one or many different responses.

9.1 Overview of Cell Communication

Recall It

Cells communicate with one another through a process called signal transduction. Signal transduction requires signal molecules, called ligands, binding to specific receptor proteins. There are various types of signaling involved in cell to cell communication, including direct signaling, paracrine signaling, endocrine signaling, and chemical synapse signaling. Molecules on the plasma membrane of one cell contact the receptor molecules on an adjacent cell in direct signaling. Paracrine signaling is where short-lived signal molecules are released into the extracellular fluid and influence neighboring cells. Endocrine signaling involves long-lived hormones that enter the circulatory system and are carried to target cells some distance away. Chemical synapse signaling uses short-lived neurotransmitters released by neurons into the gap, called a synapse, to pass messages between nerves and target cells. The phosphorylation–dephosphorylation of proteins is a common mechanism of controlling protein function.

Essential Knowledge covered
3.D.1: Cell communication processes share common features that reflect a shared evolutionary history.
3.D.2: Cells communicate with each other through direct contact with other cells or from a distance via chemical signaling.

Review It

Identify the most probable type of signaling that would be used in the following Example:

Example	Signal
A signal is sent from one cell in the body to another cell very far away	endocrine
Two cells close together communicate during early development	paracrine
Your brain perceives the pan you just picked up is hot	synaptic

What is the difference between protein kinases and phosphatases?

protein kinase is the class of enzymes adding phosphate groups from ATP to proteins and phosphates removes phosphate groups reversing the phosporalation as a result

Use It

You are taking a walk outside in a garden and almost step on a big snake that slithers in front of you. Startled, you shriek and jump! After it vanishes under a rock, you realize that your heart is pounding and you are breathing heavily. Why was your body able to react to the snake in the way it did? Think about the different types of cell signals you read about in this section.

Your heart is beating faster and you are more alert due to your body releasing hormones epinephrine raise/adrenaline

Using a diagram, show how the secretions from one cell may have an effect cells in that immediate area. What is this type of signaling known as?

synaptic signaling

neurotransmitters released

target cell

Recall It

In order for a cell to respond to a specific signal molecule, a receptor molecule is needed. Receptors are defined by being either membrane or intracellular receptors. Membrane receptors transfer information across the membrane into the cell. Membrane receptors include three subclasses: (1) channel-linked receptors, (2) enzymatic receptors, and (3) G-protein-coupled receptors. Channel-linked receptors are chemically gated ion channels that allow specific ions to pass through. Enzymatic receptors are enzymes (usually protein kinases) activated by binding a ligand. G protein–coupled receptors interact with G proteins that control the function of enzymes or ion channels. Some membrane receptors generate second messengers.

Essential Knowledge covered
3.D.3: Signal transduction pathway link signal reception with cellular response.

Review It

Identify the cell receptor type based on its structure

Receptor Type	Structure
Enzymatic receptor	A single pass transmembrane protein that catalyzes a response intracellularly
G-protein – coupled reaction	Has a binding site for a G-protein
Chemically gated ion channel	A transmembrane protein that forms a central pore; it is triggered to open or close

Use It

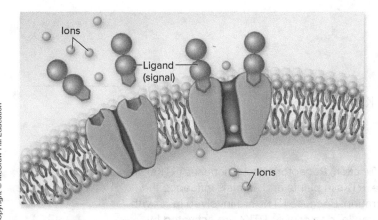

This diagram shows one of the cell receptors you learned about in this section. Correctly identify the receptor, and explain how it works.

The chemically gated ion channels form a pore in the plasma membrane that can be opened or closed by chemical signals. They are receptor types that are most of the time selective.

Recall It

Intracellular receptors are located in the cytoplasm or nucleus of a cell. Intracellular receptors interact with lipid-soluble messenger molecules which readily pass through the plasma membrane. Steroid hormones are an important group of intracellular messengers. Steroid hormones bind to receptors in the cytoplasm, and the hormone-receptor complex moves to the nucleus. These hormone-receptor complexes, also known as nuclear receptors, can then directly affect gene expression, usually activating transcription of the genes they control. A cell's response to a lipid-soluble signal depends on the hormone–receptor complex and the other protein coactivators present.

Essential Knowledge covered
3.B.2: A variety of intercellular and intracellular signal transmission mediate gene expression.

Review It

Name two different types of steroid hormones.

List two places intracellular receptors might be found.

Use It

Create a drawing of a steroid hormone-receptor complex that is ready to enter the nucleus and change gene expression.

Recall It

Protein kinases phosphorylate proteins and alter protein function. Receptor tyrosine kinases (RTKs) are receptors that can phosphorylate the amino acid tyrosine. RTKs in plants and animals influence the cell cycle, cell migration, cell metabolism, and cell proliferation. Because RTKs are involved in growth control, alterations of RTKs and their signaling pathways can lead to cancer. RTKs are activated by autophosphorylation and can also phosphorylate other intracellular proteins, activate enzymes, creating a cascade of signals throughout the cell. After producing this cascade signaling effect, internalized receptors are degraded or recycled. Small proteins act as molecular switches linking external signals to signal transduction pathways and then the RTKs are inactivated by internalization.

Essential Knowledge covered
3.B.2: A variety of intercellular and intracellular signal transmission mediate gene expression.
3.D.1: Cell communication processes share common features that reflect a shared evolutionary history.
3.D.3: Signal transduction pathways link signal reception with cellular response.

Review It

Define autophosphorylation.

when they can add phosphastes by themselves An event that transmitts across the membrane. Then propagation of the signal in the cytoplasm make a variety of diff forms.

What is a phosphorylation cascade?

series of protein kinease that phosphoralate each other succession. The final is the activation by phosphorylating

Use It

The insulin receptor is a RTK. It can mediate a number of responses, including activating a pathway that ultimately leads to the conversion of glucose to glycogen. Given what you know about RTKs and signal transduction pathways, describe some structures and processes that are most likely involved in this pathway.

RTK's & transduction pathways influence growth in cell cycle, cell migration, cell metabolism and cell proliferation ions.

Recall It

G protein–coupled receptors (GPCRs) make up the largest category of receptor types in animal cells. GPCRs bind to diverse ligands, including ions, organic odorants, peptides, proteins, and lipids. The G protein in the GPCR acts as a link between the receptor that receives a signal and another protein called an "effector" protein (usually an enzyme). The G protein can be switched on or off by the receptor. When the G protein is switched on, it activates the effector protein and causes a cellular response. The effector protein can directly cause a cellular response, but more often produces multiple second messengers which allow for further activation of proteins in a signal transduction pathway. Second messengers include molecules such as cAMP, Ca^{2+}, and IP_3. GPCRs can activate the same pathways as RTKs.

Essential Knowledge covered
3.D.3: Signal transduction pathway link signal reception with cellular response.
3.D.4: Changes in signal transduction pathways can alter cellular response.

Review It

Name the three parts that a GPCR needs to send a signal from an incoming signal molecule to the rest of the cell.

Identify the following second messengers.

Description	Secondary Messenger
This messenger is produced when a GPCR that uses the effector protein adenylyl cyclase is activated and, in turn, can activate protein kinase A and allow for phosphorylations to occur.	
This messenger has lipid ends and binds to the substrate of the effector protein phospholipase C. It is often linked to the regulation of Ca^{2+}.	
This messenger can invoke a wide range of cellular responses, including muscle contractions and hormone secretions. It is often initiates cellular responses by binding to the protein calmodulin.	

Use It

The diagram below shows a signal transduction pathway covered in this section. Fill out the table identifying the different parts of the pathway, and then, in your own words describe what is happening in the pathway.

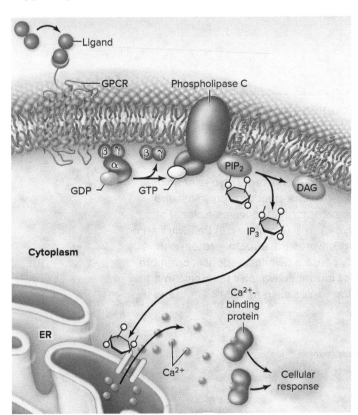

Receptor-type	
First activated protein in pathway	
Secondary messengers released	
Enzymes	
Protein that causes cellular response	

Summarize It

1. The hormone cortisol was found to enter a nucleus of a liver cell and activate the process of gluconeogenesis – the conversion of proteins and fat into glucose. Construct an explanation of how cortisol may function.

2. Construct a rational explanation of why the enzyme protein kinase is found in so many different organisms - from yeast to worms to humans.

3. A neuron releases the molecule acetylcholine. Acetylcholine receptors are found on muscle cell membranes, which open Na^+ channels when activate. The flow of Na^+ into a cell is ultimately responsible for a muscle contraction. Given this information, construct a model depicting how a neuron may cause a muscle contraction.

4. How can a hydrophilic molecule that is found only in the extracellular matrix and cannot cross the plasma membrane cause a protein to become active inside a cell?

5. Ras is a G protein that links two important different types of external and internal signal pathways. Ras is switched on when bound to the molecule GTP and switched off when bound to GDP. Ras is in a class of G proteins that can affect cell growth and differentiation. If Ras was mutated so that it was permanently stuck in an "on" positon, what effect might this have?

Chapter 10: How Cells Divide

Essential Knowledge

1.B.1	Organisms share many conserved core processes and features that evolved and are widely distributed among organisms today. **(10.2)**	**Big Idea 1**
3.A.1	DNA, and in some cases RNA, is the primary source of heritable information. **(10.1)**	**Big Idea 3**
3.A.2	In eukaryotes, heritable information is passed to the next generation via processes that include the cell cycle and mitosis or meiosis plus fertilization. **(10.3, 10.4, 10.6)**	**Big Idea 3**

Chapter Overview

All organisms grow and reproduce. Prokaryotes and eukaryotes have different forms of cell division. Prokaryotes divide through binary fission, whereas eukaryotes undergo a more complex process in order to replicate their linear chromosomes.

10.1 Bacterial Cells

Recall It

Bacteria reproduce through a simple form of cell division called binary fission. Binary fission is a type of cell division which results in two identical cells. In bacteria, DNA replication and partitioning of the chromosomes occur simultaneously. DNA replication begins at a specific point called the origin of replication and proceeds bidirectionally to a specific termination site. Newly replicated chromosomes segregate to opposite poles at the same time as they are replicated. New cells are separated by septation, which is process that involves insertion of new cell membrane and other cellular materials at the midpoint of the cell pinching the cell into two new cells.

Essential Knowledge covered
3.A.1: DNA, and in some cases RNA, is the primary source of heritable information.

Review It

Describe the steps of binary fission as illustrated in the following diagram.

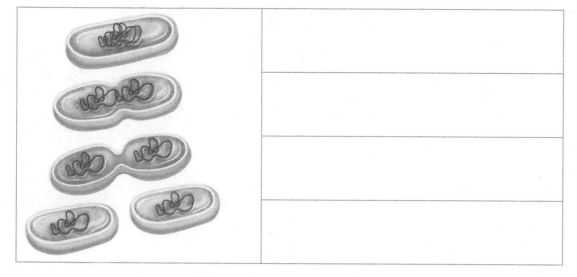

Use It

Explain why the processes of DNA replication and cell division are simpler in bacterial cells than in eukaryotic cells.

[handwritten] is† chromosome separation occurs later. If DNA in bacterial cells only have mitosis and cytokinesis in Eukaryotic cells it has to go through binary fusion.

Why is DNA replicated during binary fission?

[handwritten] to form 2 DNA molecules, which when it later divides to form daughter cells. After division the cell grows and the process repeats

Recall It

Eukaryotic chromosomes have a complex structure. They are composed of chromatin, a complex of DNA, and proteins. The DNA of a single chromosome forms a very long double-stranded fiber. In order to fit in a cell, DNA needs to be compacted. Before cell division, DNA is wrapped around a core of histone proteins to form a nucleosome, which can be further coiled. After chromosomes are replicated, they remain attached at a constricted area called a centromere. After replication, a chromosome consists of two sister chromatids held together at the centromere. Some regions of chromatin, called heterochromatin, are not expressed, while regions called euchromatin are expressed. The number of chromosome an organism has varies by species. Genetics refer to the number of different chromosomes in a species as its haploid number (n), and the totally number of chromosomes in a cell as its diploid number ($2n$).

Essential Knowledge covered
1.B.1: Organisms share many conserved core processes and features that evolved and are widely distributed among organisms today.

Review It

Provide the definition for the following parts of a eukaryotic chromosome.

Structure	Definition
Chromatin	complex of DNA protein
Histone	portion of e. chromosome not transcribed in RNA
Euchromatin	E. chromosome portion not transcribed in mRNA
Nucleosome	DNA complex consisting of DNA duplex wound
Centromere	visible point of construction of chromosome having repeated DNA sequence
Sister chromatids	1 of 2 identical copies of each chromosome still linked @ centromere
Heterochromatin	any 8 protein w/ overall positive charge
Kinetochore	Disk shaped protein structure within centromere

Humans have a haploid number of 23. What is our diploid number, and what does this mean?

our diploid # is 46 reflecting equal genetic contribution that each parent gives to the offspring.

Use It

Mosquitos have 6 chromosomes, dogs have 78, and an Adder's tongue fern has 1262 chromosomes. Given this information, what can we infer about the relationship between these species? It's an evolution of each one with different back ground evolution.

Recall It

The eukaryotic cell cycle is divided into five phases: Gap 1 (G_1), synthesis (S), gap 2 (G_2), mitosis, and cytokinesis. Together, G_1, S, and G_2 are referred to as interphase, the portion of the cell cycle necessary for growth, DNA synthesis, and preparation for division. Mitosis is subdivided into five stages, which cover the processes of DNA replication and separation of the two daughter genomes. Cytokinesis is the phase in which the cytoplasm divides, creating two cells. Mitosis and cytokinesis together are called M phase. The length of a cell cycle varies with age, cell type, and species. Cells can exit G_1 and enter a non-dividing phase called G_0, which can be temporary or permanent.

Essential Knowledge covered
3.A.2: In eukaryotes, heritable information is passed to the next generation via processes that include the cell cycle and mitosis or meiosis plus fertilization.

Review It

Describe what happens in the phases of the cycle cells as depicted in the diagram below.

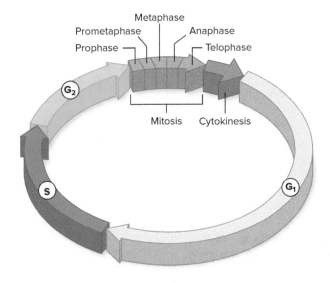

G_1 - primary growth phase

S - cell synthesizes a replica of the genome

G_2 - 2nd growth phase, and prep. for sep. of new genome

Mitosis - phase where cell cycle in which spindle assembles, binds to chromosomes and move sis. chromatids apart

Cytokin. - 2 new daughter cells are made

Use It

A cell pauses after G_1 phase and rests before entering DNA replication in a state called G_0. How long may this cell remain in this state? Can it ever re-enter the cell cycle?

The cell can remain in G_0 state for days to years before resuming cell div.

Recall It

Interphase is broken down into three subphases: G_1, S, and G_2. G_1 is the primary growth phase. During S phase, DNA synthesis occurs. During G_2 phase, chromosomes coil tightly and cells begin to assemble machinery they will later use to separate chromosomes. Centromeric DNA is replicated, but the two DNA strands are held together by cohesin proteins.

Essential Knowledge covered
3.A.2: In eukaryotes, heritable information is passed to the next generation via processes that include the cell cycle and mitosis or meiosis plus fertilization.

Review It

Identify the phases of interphase and place them in order of occurrence (1–3).

Subphase	Order of Occurrence	Description
$\mathcal{A}$	2	Chromosomes replicate
G_1	1	Cell undergo major portion of their growth
G_2	3	Animal cells complete assembly of centrioles

Use It

How does interphase prepare a cell for mitosis?

Because they went through periods of active growth where proteins are made and cell organelles are made

Recall It

M phase is divided in mitosis and cytokinesis. Mitosis is further subdivided into five stages: (1) prophase, (2) prometaphase, (3) metaphase, (4) anaphase, and (5) telophase. During prophase, chromosomes condense, and the nuclear envelope disintegrates. During prometaphase, chromosomes attach to microtubules at their kinetochores. During metaphase, chromosomes align at the equator. In anaphase, the chromatids of each chromosome are pulled to opposite poles by kinetochore microtubules. During telophase, the nucleus re-forms. Telophase reverses the events of prophase and prepares the cell for cytokinesis. Cytokinesis is the phase in which the cytoplasm divides, creating two cells. In animal cells, a belt of actin pinches off the daughter cells. In plant cells, a cell plate divides the daughter cells. In fungi and some protists, daughter nuclei are separated during cytokinesis.

Review It

Starting with a 2n = 4 cell, draw the stages of mitosis.

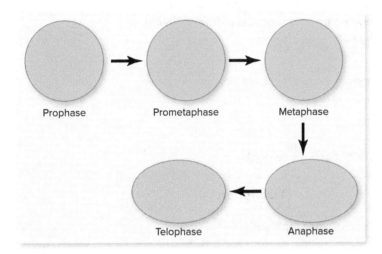

If a cell (2n = 12) underwent mitosis, what would be the following outcome?

    a. Four identical cells (2n = 6)

    b. Two identical cells (2n = 6)

    c. Two identical cells (2n = 12)

    d. Four identical cells (2n = 12)

Describe two ways in which mitosis plays a role in the life of an organism.

Why is S phase in interphase a critical step to occur before mitosis?

Recall It

The cell cycle is controlled at specific points called checkpoints. Checkpoints allow the cell to assess the accuracy of the process and stop if needed. The cell cycle is controlled by three checkpoints: the G_1/S checkpoint, the G_2/M checkpoint, and the spindle checkpoint. At the G_1/S checkpoint, the cell make a commitment to divide. The G_2/M checkpoint ensures DNA integrity, and the spindle checkpoint ensures that all chromosomes are attached to spindle fibers, with bipolar orientation. There are proteins produced in synchrony with the cell cycle called cyclins. Cyclin is required to activate regulators of mitosis, called cyclin-dependent kinases (Cdks) which phosphorylate other proteins to drive the cell cycle. Cdks and kinases active at the G_2/M checkpoint are known as M phase-promoting factor (MPF). Disruptions in cell control can lead to cancer. Two basic mechanism drive cancer: mutations in proto-oncogenes that cause dominant, gain-of-function effects, and mutations in tumor-suppressor that cause loss of function, allowing cells to grow uncontrollably.

Essential Knowledge covered
3.A.2: In eukaryotes, heritable information is passed to the next generation via processes that include the cell cycle and mitosis or meiosis plus fertilization.

Review It

Provide a description of each cell checkpoint and describe when it occurs during the cell cycle.

Checkpoint	Description	Timing in Cell Cycle
G_1/S checkpoint	where external signals influence later events	End of G_1
G_2/M checkpoint	important for stimulating events in mitosis	End of G_2
The spindle checkpoint	ensures all chromosomes are attached to spindle for anaphase	Metaphase

Draw a picture of how cyclin and Cdk interact.

Use It

Describe the importance of MPF at the G_2/M checkpoint. How does it regulate the cell cycle?

respresents the commitment to mitosis. assesses sucess of DNA replication and can the cycle if DNA hasnt been accurately copied.

The Rb protein is necessary for preventing cell growth in the retina of the eye. Rb inhibits the cell cycle by binding proteins necessary for preventing the production of cyclins. The binding power of Rb to other proteins is controlled by phosphorylation. Mutant forms of this protein can be formed by individuals with the retinoblastoma susceptibility gene (*Rb*). These individuals are predisposed to a rare form of retinal cancer. What category of cancer genes would you place *Rb*? What type of mutation do you think the *Rb* gene carries?

Must carry a gene responsible for retinoblastoma. It is also a tumor suppressor gene meaning it regulates normally.

10 | Chapter Review

Summarize It

1. The prokaryotic protein FtsZ has a structure that is similar to the eukaryotic protein tubulin, the protein component found in microtubules. In prokaryotes, FtsZ aids in septation and cell division. What role does tublin play in eukaryotic cell divison?

2. Humans have 23 pairs of chromosomes, however, humans that are missing even one chromosome generally do not survive embryonic development. Why is that one chromosome critical when there are 45 more?

That one chromosomes is a sex one most likely and with that they must have no gender and will not be able to survive as a male or female

3. The following graph shows the importance of MPF activity and cyclin protein during the stages of the cell cycle.

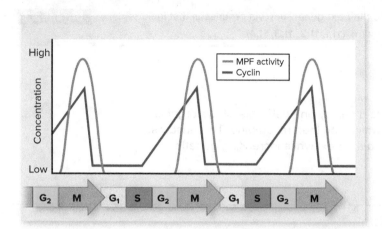

Explain why cyclin increases before MPF activity, and why MPF activity is only active during M and not G_1, S, or G_2.

From the Graph you can tell that MPF activity is only active during M and not the others because that part corresponds to the graphs in that point/peak.

Chapter 11: Sexual Reproduction and Meiosis

Essential Knowledge

3.A.2	In eukaryotes, heritable information is passed to the next generation via processes that include the cell cycle and mitosis or meiosis plus fertilization. **(11.1, 11.2, 11.3, 11.4)**	**Big Idea 3**

Chapter Overview

Meiosis reduces the number of chromosomes in a diploid cell by one half. This allows organisms that reproduce sexually to produce haploid gametes, which can then recombine. This ensures a consistent chromosome number from one generation to the next while increasing genetic variation in offspring.

11.1   Sexual Reproduction Requires Meiosis

Recall It

Meiosis reduces the number of chromosomes of diploid cells by one-half. Cells that eventually will form haploid gametes by meiosis are called germ-line cells. Germ-line cells are set aside early in animal development. Eggs and sperm are haploid gametes, and fuse during sexual reproduction to result in a diploid ($2n$) zygote. Somatic cells are diploid as well, but only ever divide by mitosis to form the body of an organism. They do not produce gametes.

Essential Knowledge covered
3.A.2: In eukaryotes, heritable information is passed to the next generation via processes that include the cell cycle and mitosis or meiosis plus fertilization.

Review It

Determine if the following statement applies to a zygote **(Z)** or a gamete **(G)**

Composed of diploid cells

Produced from germ-line cells from meiotic division

Composed of haploid cells

Undergoes mitosis to form germ-line cells and somatic cells

Use It

If a cell with a diploid number of 14 undergoes meiosis, how many chromosomes will be in the resulting gamete cells?

Offspring that inherit chromosomes from two parents are the result of what type of fertilization?

Recall It

Meiotic cell division consists of two rounds of division: meiosis I and meiosis II. During meiosis I, homologous chromosomes pair up in a process called synapsis. During synapsis, chromosomal material is exchanged via the crossing over between homologous chromosomes. At metaphase I, the paired homologues move as a unit to the metaphase plate. This causes the homologues of each pair to be pulled to opposite poles during anaphase I. At the end of meiosis I, two haploid cells are produced. The two haploid cells undergo meiosis II, which is much like mitosis but without replication of DNA. The result of meiosis II is four haploid cells.

Essential Knowledge covered
3.A.2: In eukaryotes, heritable information is passed to the next generation via processes that include the cell cycle and mitosis or meiosis plus fertilization.

Review It

Label the cells as either diploid (2*n*) or haploid (1*n*).

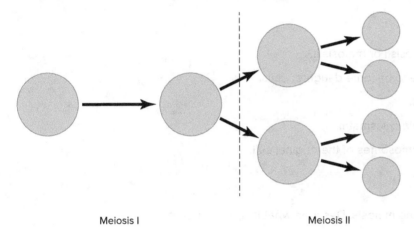

Meiosis I   Meiosis II

Use It

Describe two differences between meiosis and mitosis.

Compare and contrast meiosis I and meiosis II.

Recall It

Similar to mitosis, meiotic cells have an interphase period with G_1, S, and G_2 phases. During prophase I, chromosomes are brought closely together and aligned side by side. Sister chromatids are held together by cohesin proteins. Chromosomal material is then exchanged by crossing over. Crossing over also assists in holding the homologues together during meiosis I. Paired homologues align during metaphase. The kinetochores of sister chromatids behave as a single unit, so that homologues pairs move as a unit. During anaphase I, homologues of each pair are pulled to opposite poles as kinetochore microtubules shorten. Homologues separate, but sister chromatids to stay together. During telophase I the nuclear envelope re-forms around each daughter nucleus. Meiosis II begins the second stage of meiosis, aligning sister chromatids and producing four haploid cells that are not identical. Occasionally, errors can occurring during meiosis producing gametes with an improper number of chromosomes. This can lead to genetic disorders or even death.

Essential Knowledge covered
3.A.2: In eukaryotes, heritable information is passed to the next generation via processes that include the cell cycle and mitosis or meiosis plus fertilization.

Review It

Identify if the following statement applies to meiosis I or meiosis II.

During anaphase, sister chromatids separate to become daughter cells.

Forms four genetically dissimilar haploid cells.

Homologous chromosome pairs are linked via chiasmata.

Forms two cells with half the number of chromosomes of the original cell.

Use It

Below is a drawing of a process that occurs during meiosis. Describe what it is, and when it takes place.

 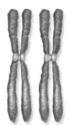

Describe two important aspects of the process illustrated above.

Recall It

While the basic machinery of chromosome movement in meiosis and mitosis is the same, the way the chromosomes move make the outcome of the processes different. Four features make meiosis distinct from in mitosis. First, maternal and paternal homologues pair during meiosis I, and exchange genetic information by crossing over. Second, sister chromatids remain connected during meiosis I and segregate together during anaphase I. Third, kinetochores of sister chromatids are connected to a single pole in meiosis I and to opposite poles in mitosis. Fourth, DNA replication is suppressed between meiosis I and meiosis II. Meiosis is critical for the process of sexual reproduction and genetic variation.

Essential Knowledge covered
3.A.2: In eukaryotes, heritable information is passed to the next generation via processes that include the cell cycle and mitosis or meiosis plus fertilization.

Review It

Identify if the following statement applies to meiosis or mitosis.

Sister kinetochores are attached to the same pole.

Homologous pairing is specific to this process.

This process ensures genetic diversity in sexually reproducing organisms.

This processes produces identical cells.

Use It

Why does meiosis result in a reduction of chromosome number while mitosis produces daughter cells with the same chromosome number?

11 Chapter Review

Summarize It

A cell with a diploid number of 4 undergoes meiosis. Using a drawing, describe the stages of meiosis and the result of the process.

Chapter 12: Patterns of Inheritance

Essential Knowledge

3.A.3	The chromosomal basis of inheritance provides an understanding of the patterns of passage (transmission) of genes from parent to offspring. **(12.1, 12.2, 12.3, 12.4)**	**Big Idea 3**
3.A.4	The inheritance pattern of many traits cannot be explained by simple Mendelian genetics. **(12.6)**	**Big Idea 3**

Chapter Overview

Gregor Mendel's experiments with pea plants led to our understanding of how genes are passed from one generation to the next. Mendel identified that there are different forms, or alleles, for genes and that these alleles separate during gamete formation. Different traits can segregate independently of one another. Most often, probability can be used to determine the outcome of inheritance, but the inheritance of some traits are harder to predict for various reasons, including incomplete dominance, polygenic inheritance, pleiotropy, and environmental effect on phenotype.

12.1 The Mystery of Heredity

Recall It

The concept of inheritance was not well understood until Gregor Mendel conducted a series of experiments in the late 1850s. Unlike early plant biologists who were trying to demonstrate the blending theory qualitatively, Mendel was the first to quantify the results of his crosses. Plant breeders, including Mendel, noticed that some forms of a trait could disappear in one generation only to reappear later. Mendel realized that these traits segregate rather than blend. Mendel's experiments involved cross breeding between true-breeding pea varieties (plants that produced uniform offspring) followed by one or more generations of self-fertilization. Mendel's mathematical analysis of experimental results led to the present model of inheritance.

Essential Knowledge covered
3.A.3: The chromosomal basis of inheritance provides an understanding of the pattern of passage (transmission) of genes from parents to offspring.

Review It

List three reasons Mendel chose the garden pea plant on which to conduct his experiments.

Use It

Describe two ways in which Mendel's pollination work was different from earlier plant biologists.

Mendel used plants that were self-fertilizing, but he could manipulate the plants to be cross-pollinated as well. Why was it important to his study that Mendel knew where the pollen was coming from to fertilize his plants?

12.2 Monohybrid Crosses: The Principle of Segregation

Recall It

A monohybrid cross is a cross that allows for only two variations on a single trait. Mendel called the trait expressed in the F_1 generation the dominant trait; the other he called recessive. Mendel allowed F_1 plants to mature and self-fertilize, and found that the F_2 generation exhibited a 3:1 ratio of dominant:recessive traits. Mendel then allowed the F_2 generation to self-fertilize and found the recessive F_2 plants always bred true, but only one out of three dominant F_2 bred true. The 3:1 ratio was actually a disguised 1:2:1. We call this ratio the Mendelian monohybrid ratio. Mendel's Principle of Segregation explains this observation. The Principle of Segregation states that during gamete formation, the two alleles of a gene segregate into different gametes. Individuals carrying two identical alleles for a gene are said to be homozygous, and individuals carrying different alleles are said to be heterozygous. An individual's genotype is the entire set of alleles of all genes possessed by an individual. The individual's appearance or phenotype is due to which alleles are expressed.

Essential Knowledge covered
3.A.3: The chromosomal basis of inheritance provides an understanding of the pattern of passage (transmission) of genes from parents to offspring.

Review It

Define the following vocabulary words that pertain to the field of genetics.

Word	Definition
Genotype	
F_1 generation	
Domain	
Heterozygous	
Phenotype	
Recessive	
Allele	
Homozygous	
Pedigree	

Use It

In a certain pea plant, yellow peas (Y) are dominant over green peas (y). If a cross is done between a plant with yellow peas (YY) and a plant with green peas (yy), what color peas will the F_1 generation exhibit?

If the F_2 generation of the plants from the question above were allowed to self-fertilize, describe the genotypic and phenotypic ratios of the resulting F_2 generation's offspring. Use a Punnett square to help arrive at your answer.

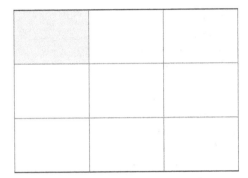

12.3 Dihybrid Crosses: The Principle of Independent Assortment

Recall It

Dihybrid crosses follow the inheritance of two different traits in a single cross. The two traits in a dihybrid cross behave independently. If a parent with two homozygous dominant traits (XXYY) is crossed with a parent with two homozygous recessive (xxyy) traits, all F_1 individuals will have the same genotype and phenotype (XxYy). Each F_1 parent can then produce four different gametes that can be combined in the F_2 generation. This yields a phenotypic ratio of 9:3:3:1 of the four possible phenotypes. Mendel's Principle of Independent Assortment explains these dihybrid results, stating that different traits segregate independently of one another. The physical basis of independent assortment is the independent behavior of different pairs of homologous chromosomes during meiosis.

Essential Knowledge covered
3.A.3: The chromosomal basis of inheritance provides an understanding of the pattern of passage (transmission) of genes from parents to offspring.

Review It

In the dihybrid cross below, list all the different possible phenotypic combinations and the ratio of each, if R is dominant for round and recessive for wrinkled (r), and Y is dominant for yellow and recessive for green (y).

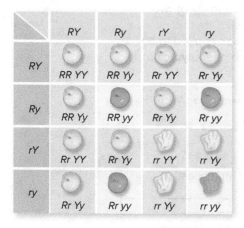

Use It

Independently assort these two homologous chromosomes into the four gamete cells.

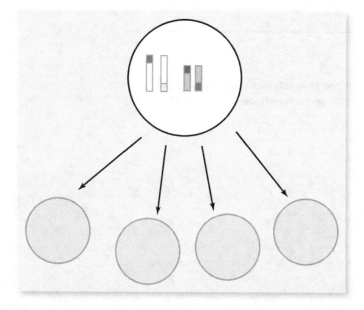

If one set of the chromosomes in the parent cell above carried alleles for the traits purple flowers (Pp) and one set carried alleles for tall plant height (Tt) and the plant self-fertilized, what would the resulting genotype and phenotypes be in the daughter cells?

Recall It

The likelihood of an outcome of a random event can be predicted by probability. Two probability rules help predict monohybrid cross results: (1) the rule of addition and (2) the rule of multiplication. The rule of addition states that the probability of two independent events occurring is the sum of their individual probabilities. The rule of multiplication states that the probability of two independent events both occurring is the product of their individual probabilities. Dihybrid cross probabilities can be determined from monohybrid cross probabilities. Ratios applied to an observed total of offspring can be used to predict the expected number in each phenotypic group.

Essential Knowledge covered
3.A.3: The chromosomal basis of inheritance provides an understanding of the pattern of passage (transmission) of genes from parents to offspring.

Review It

Determine if the following is either an example of the rule of addition or rule of multiplication.

The probability of F_2 heterozygote from a heterozygous purple F_1	The probability of a *pp* homozygote from a heterozygous male parent and a heterozygous female parent

Use It

Using the two probability rules, calculate the probabilities for the phenotypes of a dihybrid cross between an individual with purple flowers and big leaves (PpLl) and white flowers and small leaves (ppll).

Recall It

Mendel devised a procedure called the testcross to reveal an unknown genotype of an offspring. He did this by crossing the unknown with a homozygous recessive genotype. Mendel realized he could ignore the contribution of the recessive genotype and the resulting phenotype of the unknown would illuminate its genotype.

Review It

Consider the following flower.

Dominant Phenotype (unknown genotype)

PP or *Pp*

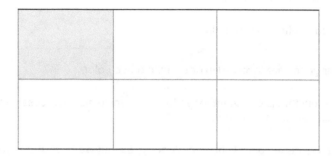

It is purple, but its genotype is unknown. If it is crossed with a homozygous recessive (pp), and half of the offspring have white flowers and half of the offspring have purple flowers, what was its genotype? Show your results in the Punnett Square provided for you above.

12.6 Extensions to Mendel

Recall It

As some traits are the product of many genes and/or physiological patterns, the inheritance pattern of these traits cannot be explained by simple Mendelian genetics. More than one gene can affect a single trait (polygenic inheritance), a single gene can affect more than one trait (pleiotropy), or a gene have may have more than two alleles. Sometimes, a dominant trait is not always completely expressed, and a heterozygote exhibits an intermediate phenotype even though the monohybrid genotypic and phenotypic ratios are the same. Phenotypes may also be affected by environmental conditions, both external and internal factors. Finally, genetic ratios may be effected by the product of other genes.

Essential Knowledge covered
3.A.4: The inheritance pattern of many traits cannot be explained by simple Mendelian genetics.

Review It

Identify the type of genetic occurrence based on the example.

Genetic Occurrence	Example
	The allele that produces dark pigment in Himalayan rabbits is heat-sensitive and becomes inactive at high temperatures.
	Four o'clock flowers that contain an allele for red flowers and an allele for white flowers have pink flowers.
	Epistasis affects corn grain color.
	Four genes are involved in determining eye color.
	An allele that is dominant for yellow fur in mice also is recessive for a lethal developmental defect.
	A single gene controls human blood types, but we have ABO groups.

Use It

Explain why human height is a continuously varying trait.

12 Chapter Review

Summarize It

1. Using the two probability rules, calculate the probabilities for the phenotypes of a dihybrid cross between an individual with yellow flowers and red seeds (YY Rr) and white flowers and white seeds (yy rr).

2. The petals of a *Camellia* plant were found to be striped white and red. The plant was the offspring of a plant with red flowers (R) and white (r). Explain how the offspring had striped petals and not red petals.

Chapter 13: Chromosomes, Mapping, and Meiosis-Inheritance Connection

Essential Knowledge

3.A.3	The chromosomal basis of inheritance provides an understanding of the patterns of passage (transmission) of genes from parent to offspring. **(13.4, 13.5)**	**Big Idea 3**
3.A.4	The inheritance pattern of many traits cannot be explained by simple Mendelian genetics. **(13.1, 13.2, 13.3)**	**Big Idea 3**

Chapter Overview

Observing the behavior of chromosomes during meiosis has led to new and different approaches to the study of heredity. Some genes are only found on sex-linked chromosomes, and sometimes genes are found in the genomes of organelles. There are many tools available to help geneticists determine the location of genes on chromosomes and which genes may be involved in genetic diseases.

13.1 Sex Linkage and the Chromosomal Theory of Inheritance

Recall It

The chromosomal theory of inheritance states that traits are carried on chromosomes. The X and Y chromosomes are known as sex chromosomes. Traits determined by genes on the X chromosome are termed sex-linked inheritance because it is associated with the sex of the individual. This was first clearly tested by Thomas Morgan, who crossed red-eyed and white-eyed *Drosophila* flies and found differences in inheritance based on the sex of offspring. All white-eyed offspring were males, but testcrosses showed that white-eyed females were possible, supporting the idea that the white-eye gene was on the X chromosome.

Essential Knowledge covered
3.A.4: The inheritance pattern of many traits cannot be explained by simple Mendelian genetics.

Review It

Circle the sex chromosomes.

Chromosome 13   Y Chromosome   Chromosome 2   X Chromosome

What type of experiment did Morgan perform to determine the viability of white-eyed female flies?

Use It

A woman who had worked in an animal shelter for many years one day realized that all of the calico cats (patchy orange and black fur) that came to the facility were also females. She had seen many male cats that had orange fur or black fur but never a calico male. Explain this phenomenon in terms of chromosomal inheritance.

Recall It

In many animals, sex determination is associated with a chromosomal difference. In some animals, females have two similar sex chromosomes (XX) and males have sex chromosomes that differ (XY). In other species, females have sex chromosomes that differ (ZW) or females may have two similar sex chromosomes (XX) and males only have one (XO). In humans, the Y chromosome generally determines "maleness;" any individual with at least one Y chromosome is normally a male. The Y chromosome is highly condensed and does not have active counterparts to most genes on the X chromosome. In females with two similar sex chromosomes (XX), one of the X chromosomes is inactivated early in development, becoming a Barr body. This inactivation is called dosage compensation, which ensures there is an equal level of expression from sex chromosomes in both males and females. These types of females are considered "genetic mosaics," as their individual cells may express different alleles depending on which X chromosome has been inactivated.

Essential Knowledge covered
3.A.4: The inheritance pattern of many traits cannot be explained by simple Mendelian genetics.

Review It

Hemophilia is a human genetic disorder that is sex-linked. What sex chromosome is the allele for hemophilia located on?

Use It

If a woman was a carrier for a sex-linked recessive baldness gene (XX^{bald}) and had children by a bald man, what percentage of their male children will be bald?

Recall It

There are few exceptions to the Chromosomal Theory of Inheritance, and these are mainly due to the presence of DNA in the organelle genomes of mitochondria and chloroplasts. Mitochondria and chloroplasts have their own genomes and divide independently. These genomes are passed to offspring in the cytoplasm of the egg cell. Traits determined from mitochondrial or chloroplast DNA will not show Mendelian inheritance patterns. Mitochondrial genes are inherited maternally, and chloroplast genes may be passed on uniparentally.

Essential Knowledge covered
3.A.4: The inheritance pattern of many traits cannot be explained by simple Mendelian genetics.

Review It

Name two organelles that have their own genomes.

How is maternal inheritance different from sex-linked inheritance?

Use It

Why do genes on the mitochondrial genome not show Mendelian inheritance patterns?

Four o'clock plants have a mixture of leaf colors; some are green and some are white. Pollen from branches with green leaves used to fertilize branches with white leaves will produce offspring with only white leaves. Similarly, pollen from branches with white leaves used to fertilize branches with green leaves will only produce plants with green leaves. Explain this inheritance pattern of leaf pigment.

Copyright © McGraw-Hill Education

13.4 Genetic Mapping

Recall It

When a crossover occurs during meiosis, it leads to the production of recombinant chromosomes. Recombinant chromosomes will result in genetically recombinant progeny. Genes that are close together on a single chromosome are referred to as linked and have a lower frequency of recombination. Genes that are far apart have a greater frequency of recombination and a greater chance of multiple crossovers. Genetic maps can be constructed based on recombination frequency, using data from genetic crosses. The frequency of recombination is defined by the number of recombinant progeny divided by total progeny. Evaluating genes with less separation, provides more accurate distances as genes separated by large distance are more likely to independently assort. Three-point crosses can be used to put genes in order. Genetic markers help to make genetic maps, such as the map of the human genome.

Essential Knowledge covered
3.A.3: The chromosomal basis of inheritance provides an understanding of the pattern of passage (transmission) of genes from parents to offspring.

Review It

What are SNPs?

Determine if the following statements regarding genetic mapping are true or false **(T/F)**:

Genes that are adjacent and close to each other on the same chromosome tend to move as a unit.

Independent assortment of chromosomes almost never results in genetic variation.

Recombination is the basis for genetic maps.

The human genome is too large to have a genetic map constructed of it.

The closer two linked genes are, the greater the frequency of recombination.

Use It

Drosophila homozygous for two mutations, curly wings (cy) and small eyes (ey), are crossed with wild type fruit flies (cy$^+$, ey$^+$). This F$_1$ generation is then testcrossed to homozygous recessive individuals, and the progeny are counted. Using the data in the chart below, map the distance between these two loci.

Phenotype	Number
Curly wings, small eyes (cy, ey)	390 (parental mutant type)
Long wings, normal eyes (cy$^+$, ey$^+$)	385 (parental wild type)
Curly wings, normal eyes (cy, ey$^+$)	110 (recombinant)
Long wings, small eyes (cy$^+$, ey)	115 (recombinant)
Total Progeny	1000

What happens to the probability of recombination occurring between genes as the distance separating loci increases?

Recall It

Genetic disorders can be caused by a change in chromosomal number or changes at the DNA level. Nondisjunction is the failure of homologues or sister chromatids to separate properly during meiosis. Nondisjunction can cause aneuploidy- the gain or loss of a chromosome. Losing a copy of a chromosome is known as monosomy, and gaining an extra copy is called trisomy. Most aneuploidies are lethal, but some, such as trisomy 21 in humans (Down syndrome), can result in viable offspring. Sex chromosome nondisjunction can produce XX, YY, or gametes with no sex chromosomes. Other genetic disorders arise from mutations or alterations of genes, which affect protein structure and function. Genetic disorders can also be the result of genomic imprinting, in which the expression of a gene depends on the parent of origin. Genomic imprinting is an example of epigenetic inheritance, in which a heritable change in phenotype is not due to a mutation in DNA sequence but a stable change, such as DNA methylation, which has led to gene inactivation.

Essential Knowledge covered
3.A.3: The chromosomal basis of inheritance provides an understanding of the pattern of passage (transmission) of genes from parents to offspring.

Review It

Using your textbook, identify the cause of the following disorders.

Genetic disorder	Cause
Tay-Sachs disease	
Trisomy 21	
Prader-Willi syndrome	
X-linked color blindness	
Huntington's diseases	
Klinefelter's syndrome	

Use It

Why would a nondisjunction event of sex chromosomes that results in an individual with Klinefelter syndrome (XXY) cause more serious defects than an individual with Triple X syndrome (XXX)?

Sickle cell anemia is a disorder in which individuals have irregular shaped red blood cells and impaired oxygen delivery to tissues. Describe what causes this disorder.

Summarize It

1. What question would you ask your doctor to determine if a medical disorder you found out you had was heritable?

2. Why are more men red-green color-blind then women?

Chapter 14: DNA: The Genetic Material

Essential Knowledge

3.A.1	DNA, and in some cases RNA, is the primary source of heritable information. **(14.1, 14.2, 14.3, 14.4, 14.5)**	**Big Idea 3**
3.C.1	Changes in genotype can result in changes in phenotype. **(14.6)**	**Big Idea 3**
3.D.1	Cell communication processes share common features that reflect a shared evolutionary history. **(14.6)**	**Big Idea 3**

Chapter Overview

DNA is the core genetic material that encodes all the instructions for life. DNA is a double stranded helix that contains a sugar-phosphate backbone, a phosphate group, and complimentary nitrogenous bases. DNA replicates in a semiconservative manner and requires many different enzymes in order to synthesize new strands successfully in both prokaryotes and eukaryotes.

14.1 The Nature of the Genetic Material

Recall It

Three major studies showed that DNA, not protein, is the source of genetic material. Griffith was the first to find that bacterial cells could be transformed. He found that nonvirulent *S. pneumoniae* bacteria could take up an unknown substance from a virulent strain and become virulent. Avery, MacLeod, and McCarty identified the substance responsible for the transformation in Griffith's experiment as DNA, by showing that DNA digesting enzymes (and not protein digesting ones) inactivated the transforming substance. Hershey and Chase confirmed these results by using radioactive labels to mark DNA and protein in bacteriophages. Hershey and Chase visually showed that the infectious agent of phage is its DNA and not its protein.

Essential Knowledge covered
3.A.1: DNA, and in some cases RNA, is the primary source of heritable information.

Review It

Place the three historical experiments discussed in this chapter on the timeline below, and provide a short description of their results.

1925   1944   1952

Use It

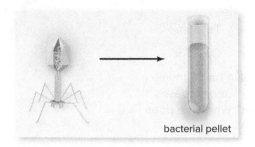

bacterial pellet

How did Hershey and Chase show that phage DNA, not protein, was responsible for infection in bacterial cells?

14.2 DNA Structure

Recall It

The three main components of DNA are a 5-carbon sugar, a phosphate group, and a nitrogen-containing base. The nucleotide base may be either adenine (A), guanine (G), cytosine (C), or thymine (T). While the nucleotide composition varied in DNA, scientist Erwin Chargaff found there were always equal amounts of adenine and thymine, and of cytosine and guanine. Phosphodiester bonds are formed between the 5′phosphate of one nucleotide and the 3′ hydroxyl of another nucleotide. X-ray diffraction studies by Franklin and Wilkins indicated that DNA had a helical structure. Watson and Crick used the results of these studies to deduce the structure of DNA. The Watson-Crick model shows two antiparallel polynucleotide strands wrapped about a common helical axis, held together by hydrogen bonds forming specific base pairs (A/T and G/C). The two strands are complementary; one strand can specify the other.

Essential Knowledge covered
3.A.1: DNA, and in some cases RNA, is the primary source of heritable information.

Review It

Using the segment of DNA on the next page, review general traits of DNA.

Draw an arrow to each phosphate group in this strand of DNA.

How many 5-carbon sugars are present in this segment of DNA?

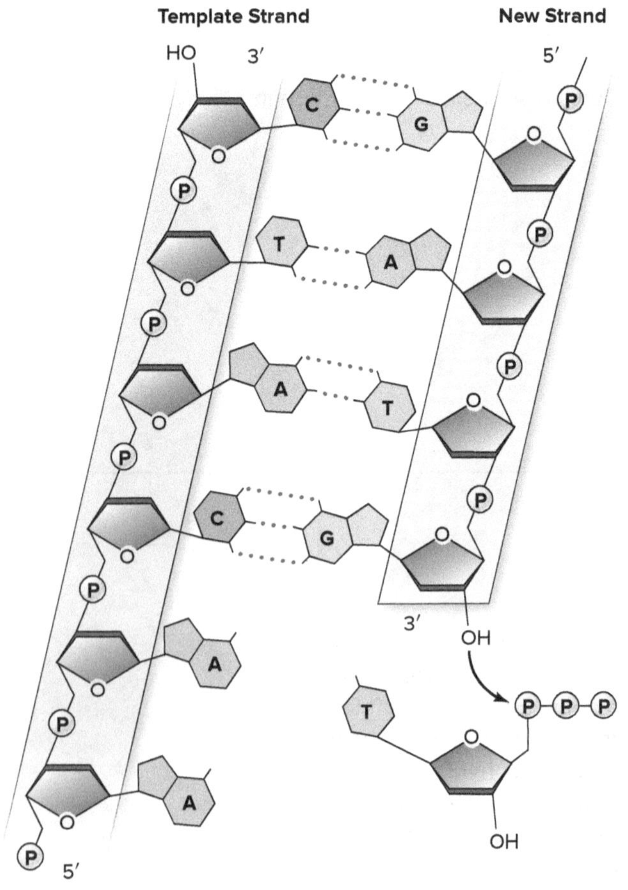

Template Strand

HO 3′

New Strand

5′

3′

OH

5′

OH

Count the number of nitrogenous bases:

Adenine: Thymine:

Guanine: Cytosine:

If a segment of DNA had 8 thymine bases, how many adenine bases would be present? What rule does this follow?

Use It

Describe the Watson-Crick model of DNA. Be sure to use and explain the terms *phosphodiester backbone, antiparallel configuration,* and *complementarity base pairs* in your answer.

The number of hydrogen bonds between a single pair of nitrogenous bases was determined to be two. What two nitrogenous bases have this pairing configuration?

Recall It

As shown through the work of Meselson and Stahl, DNA replicates in a semiconservative manner. One parental strand of a DNA remains intact in the daughter strand, while a new complementary strand is built for each parental strand. Meselson and Stahl showed this by using a heavy isotope of nitrogen and separating the replication products. DNA replication requires a template DNA strand, nucleotides, and a polymerase enzyme. All new DNA molecules are produced by DNA polymerase copying a template. The building blocks used in replication are deoxynucleotide triphosphates with high-energy bonds; they do not require any additional energy. DNA is synthesized by DNA polymerase in the 5'-3' direction.

Essential Knowledge covered
3.A.1: DNA, and in some cases RNA, is the primary source of heritable information.

Review It

Circle the correct mode of DNA replication.

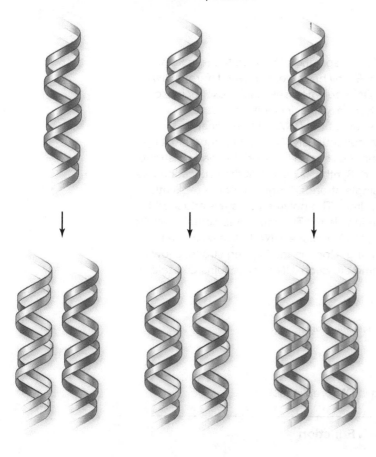

List three things DNA requires for DNA replication.

Use It

DNA replication requires DNA polymerase. What is DNA polymerase and how does it work?

DNA polymerase requires a primer to begin DNA synthesis. Where does this primer come from?

14.4 Prokaryotic Replication

Recall It

Prokaryotic replication starts at a single origin and ends of a specific site called the terminus. The DNA controlled by an origin is known as a replicon. *E. coli* has at least three different DNA polymerases that play a role in DNA replication. Some DNA polymerases, called exonuclease activity, can also remove nucleotides from one end. Other enzymes, including helicase and topoisomerase, help unwind and relieve torsional strain of the DNA. DNA replication is semidiscontinuous; only one strand can be synthesized continuously. The other strand must be made in pieces known as Okazaki fragments. The continuous strand is called the leading strand, and the discontinuous strand is called the lagging strand. Synthesis occurs at the replication fork where the partial opening of a DNA strand forms two single-stranded regions. At the replication fork, synthesis on the leading strand requires a single primer. The polymerase stays attached to the template because of the β subunit that acts as a sliding clamp. The replisome contains all the necessary enzymes for replication: two copies of Pol III, DNA primase, DNA helicase, and a number of accessory proteins. It moves in one direction by creating a loop in the lagging strand, allowing the antiparallel template strands to be copied in the same direction. DNA ligase joins Okazaki fragments into complete strands.

Essential Knowledge covered
3.A.1: DNA, and in some cases RNA, is the primary source of heritable information.

Review It

Identify the function of the following enzymes involved in DNA replication:

Enzyme	Function
DNA ligase	
DNA polymerase	
Helicase	
Topoisomerases	
RNA polymerase	

Use It

Construct a diagram showing how DNA ligase and DNA polymerase I are involved in the synthesis of a lagging strand of DNA. Be sure to identify the 5' and 3' end of the strand and any Okazaki fragments.

Are all the same enzymes involved in the production of leading and lagging strands during DNA synthesis the same ones?

14.5 Eukaryotic Replication

Recall It

Eukaryotic replication is complicated by two main factors: the sheer size and amount of DNA, and organization of eukaryotic DNA into linear chromosomes. To replicate DNA in the short time available in S phase, eukaryotic replication requires multiple origins of replication. Linear chromosomes have specialized ends of non-coding DNA called telomeres, made by the enzyme telomerase. Telomerase contains an internal RNA that acts as a template to extend the DNA of the chromosome end. Adult cells lack telomerase activity, and telomere shortening correlates with cell senescence.

Essential Knowledge covered
3.A.1: DNA, and in some cases RNA, is the primary source of heritable information.

Review It

Determine if the following statements concerning eukaryotic replication are true or false **(T/F)**:

Eukaryotic replication occurs in the exact same manner as prokaryotic replication.

Telomeres are composed by the replication complex.

In the absence of telomerase activity, the ends of chromosomes gradually become shorter.

Eukaryotic replication uses multiple origins of DNA replication for each chromosome.

Eukaryotic primase is a complex of both RNA polymerase and DNA polymerase.

Use It

Describe two ways eukaryotic DNA replication is different from prokaryotic DNA synthesis.

14.6 DNA Repair

Recall It

Cells are constantly exposed to DNA-damaging agents known as mutagens, including UV light, X-rays, and chemicals. Internal mechanisms have evolved to repair damaged DNA. Without these repair mechanisms, cells would accumulate mutations until they eventually became inviable. DNA repairs can be specific or nonspecific. Specific repair mechanisms target a single kind of lesion in DNA. For example, cells have photorepair, a repair mechanism that pinpoints and repairs a thymine dimer formed by a photochemical reactions between DNA and UV light. Nonspecific repair mechanisms target multiple kinds of lesions and repair them all at once. An example of nonspecific repair is excision repair, which removes an entire damaged region of DNA and replaces it. Nonspecific repair can also be placed into error-free or error-prone categories; an error-prone pathway most likely evolved as a last-ditch-effort to save a cell exposed to massive damage.

Essential Knowledge covered
3.C.1: Changes in genotype can result in changes in phenotype.
3.D.1: Cell communication processes share common features that reflect a shared evolutionary history.

Review It

List the three steps of excision repair.

Use It

Describe what might happen to a cell that did not have any DNA repair mechanisms.

Summarize It

Review the transformation experiment that Griffith performed (p. 257 in your textbook). How did the results of Griffith's experiment lead him to the conclusions that some type of material was responsible for the passage of heritable information?

Chapter 15: Genes and How They Work

Essential Knowledge

1.B.1	Organisms share many conserved core processes and features that evolved and are widely distributed among organisms today. **(15.1, 15.2)**	**Big Idea 1**
1.D.2	Scientific evidence from many different disciplines supports models of the origins of life. **(15.2)**	**Big Idea 1**
2.E.1	Timing and coordination of specific events are necessary for the normal development of an organism, and these events are regulated by a variety of mechanisms. **(15.4)**	**Big Idea 2**
3.A.1	DNA, and in some cases RNA, is the primary source of heritable information. **(15.1, 15.3, 15.4, 15.5, 15.7, 15.8)**	**Big Idea 3**
3.C.1	Changes in genotype can result in changes in phenotype. **(15.9)**	**Big Idea 3**
4.A.1	The subcomponents of biological molecules and their sequence determine the properties of that molecule. **(15.3)**	**Big Idea 4**
4.A.2	The structure and function of subcellular components, and their interactions, provide essential cellular processes. **(15.6)**	**Big Idea 4**

Chapter Overview

The central dogma of biology states that DNA is transcribed into RNA and translated into proteins. The genetic code is nearly universal. Prokaryotic and eukaryotic gene expression have many similar steps, but eukaryotic gene expression tends to be more complex.

15.1 The Nature of Genes

Recall It

The central dogma of molecular biology describes information flow in cells as DNA to RNA to protein. The process of transcription makes an RNA copy of DNA. Only one strand of DNA is copied (the template strand), the DNA not used as a template is called the coding strand. Translation is the process in which RNA is used to synthesize protein. There are several varieties of RNA, which have different functions. This includes messenger RNA (mRNA), ribosomal RNA (rRNA), transfer RNA (tRNA), small nuclear RNA (snRNA), signal recognition RNA (SRP RNA), and small RNA, including micro-RNA (miRNA) and small interfering RNA (siRNA). The rRNA is critical for the function of the ribosome, the tRNA interprets information in mRNA and helps to position the amino acids on the ribosome, snRNA are involved in splicing, SRP RNA mediated proteins synthesis on the rough endoplasmic reticulum in eukaryotes, and small RNAs are involved in the control of gene expression. The flow of genetic information is different in retroviruses, in which RNA is used to create a DNA copy.

Essential Knowledge covered
1.B.1: Organisms share many conserved core processes and features that evolved and are widely distributed among organisms today.
3.A.1: DNA, and in some cases RNA, is the primary source of heritable information.

Review It

Identify the RNA involved in the process:

RNA	Process
	Positions an amino acid on the ribosome
	Synthesizes protein on the rough ER
	Transports message from nucleus to cytoplasm
	Makes up ribosomal subunit
	Regulates gene expression

What is the central dogma of molecular biology?

Use It

Using the central dogma of molecular biology, describe how your DNA is used to make the protein needed in your fingernails.

Explain how the flow of genetic information is different in retroviruses than explained by the central dogma.

Recall It

The genetic code is organized into groups of three nucleotides called codons. Each codon consists of three possible nucleotides. Given that there are 4 different nucleotides, there are 4^3 or 64 possible codons. Three codons signal "stop," and one codon signals "start." The start codon also codes for the amino acid methionine. Sixty-one codons are left to encode the 20 amino acids. Many amino acids have more than one codon, but each codon specifies only a single amino acid. The genetic code is almost universal; in mitochondrial and protist genomes a stop codon is read as an amino acid.

Essential Knowledge covered
1.B.1: Organisms share many conserved core processes and features that evolved and are widely distributed among organisms today.
1.D.2: Scientific evidence from many different disciplines support models of the origin of life.

Review It

Review the numbers of the genetic code.

Number of possible codons:

Number of nucleotides in a codon:

Number of codons that indicate "start:"

Number of codons that indicate "stop:"

Number of amino acids a codon encodes:

Use It

Using Table 15.1 (p. 283 in your textbook), determine the following amino acid sequence encoded by the following codons:

AUG   UUU   CAU   CAC   AAA   UGG   UGA

Would this amino acid sequence be read the same in a human and in *E.coli*? Why or why not?

Recall It

Prokaryotes and eukaryotes have different methods of transcription. Prokaryotes have a single RNA polymerase that exists in two forms: the core polymerase and the holoenzyme. Initiation of prokaryotic transcription requires a start site and a promoter. The promoter is upstream of the start site. The first step in transcription is the binding of RNA polymerase holoenzyme to the promoter. The holoenzyme can accurately recognize the promoter sequence. Elongation follows, in which the core polymerase synthesizes RNA nucleotides. The RNA polymerase, the locally unwound DNA template, and the growing mRNA transcript are referred to as the transcription bubble. Termination occurs at a specific "stop" site and consists of complementary sequences that form a double-stranded hairpin loop where the polymerase dissociates from the DNA within the transcription bubble. In prokaryotes, translation begins while mRNAs are still being transcribed. Prokaryotic genes may be clustered together in groups known as operons.

Essential Knowledge covered
3.A.1: DNA, and in some cases RNA, is the primary source of heritable information.
4.A.1: The subcomponents of biological molecules and their sequence determine the properties of that molecule.

Review It

List the three major steps of prokaryotic transcription

Use It

How is transcription coupled to translation in prokaryotes, and why is it not coupled in eukaryotes?

Describe the direction of elongation.

Recall It

Unlike prokaryotic transcription, eukaryotes have three different RNA polymerases, which differ in structure and function. Polymerase I transcribes rRNA, polymerase II transcribes mRNA and some snRNAs, and polymerase III transcribes tRNA. Each type of polymerase has its own promoter. General transcription factors bind to the promoter and recruit polymerase II to form the initiation complex. After the transcript reaches about 20 nucleotides, it is modified by the addition of GTP to the 5′ phosphate. This modification is also known as the 5′ cap and is important for translation, RNA stability, and further processing. Additionally, a long chain of adenine residues, known as the poly-A tail, is added to the 3′ end by the enzyme poly-A polymerase. The poly-A tail protects mRNA from degradation. Finally, eukaryotic genes may have noncoding sequences called introns removed from the mRNA by an organelle called the spliceosome. Eukaryotic mRNA always has a 5′ cap, a 3′ poly-A tail, and introns that must be removed.

Essential Knowledge covered
2.E.1: Timing and coordination of specific events are necessary for the normal development of an organism, and these events are regulated by a variety of mechanisms.
3.A.1: DNA, and in some cases RNA, is the primary source of heritable information.

Review It

List two ways eukaryotic transcription is different from prokaryotic transcription.

Use It

Three posttranscriptional modifications occur to eukaryotic transcripts that are necessary for the synthesis of amino acids. What are these modifications, and how are they regulated?

Could eukaryotic transcription occur without transcription factors? Explain your answer.

Recall It

Eukaryotic genes contain noncoding DNA (introns) and coding sequencing (exons). Exons are interrupted by introns. Introns are removed after transcription by an organelle called the spliceosome. The spliceosome ultimately joins the 3' end of the first exon to the 5' end of the next exon. Splicing usually produces multiple transcripts from the same gene.

Essential Knowledge covered
3.A.1: DNA, and in some cases RNA, is the primary source of heritable information.

Review It

Define the following portion of mRNA transcript. Circle the portion that will be spliced out.

A noncoding intervening sequence

The coding sequence

Use It

Draw the mature mRNA transcript of the DNA template below:

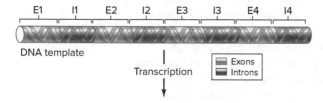

Recall It

Transfer RNA (tRNA) brings the appropriate amino acid to the mRNA chain based on the DNA code. The enzyme aminoacyl-tRNA synthetase attaches the carboxyl terminus of an amino acid to the 3' end of the correct tRNA amino acids. This reaction is known as the tRNA charging reaction. Protein synthesis takes place on the ribosome, which has multiple tRNA-binding sites. Ribosomes hold tRNAs and mRNA in position for a ribosomal enzyme to form peptide bonds.

Essential Knowledge covered
4.A.2: The structure and function of subcellular components, and their interactions, provide essential cellular processes.

Review It

List two functions of tRNA.

Use It

Create a diagram that shows and explains how tRNA interacts with both amino acids and codons.

15.7 The Process of Translation

Recall It

Translation occurs in three phases: initiation, elongation, and termination. Translation usually beings when a ribosome encounters a start codon. The start codon in eukaryotes is AUG, which encodes the amino acid methionine. In prokaryotes, a chemically modified version of methionine initiates translation. Translation needs many different accessory factors and proteins to aid ribosome binding to mRNA. During elongation, the ribosome is threaded through the mRNA, and the correct amino acids from charged tRNAs are successively added to the growing peptide. Elongation continues until a "stop" codon is reached. Like initiation, termination also requires accessory factors, in this case releasing the polypeptide chain from the ribosome. In eukaryotes, translation can occur in the cytoplasm or the rough ER. Signaling sequences help the cell recognize where a protein needs to be assembled.

Essential Knowledge covered
3.A.1: DNA, and in some cases RNA, is the primary source of heritable information.

Review It

Place the following events of translation in order (1–4).

Peptide bond formation

Matching tRNA antiocodon with mRNA codon

Termination of polypeptide

Release of polypeptide

Use It

Describe two ways translation is regulated.

15.8 Summarizing the Gene Expression

Review It

Fill in the blanks to summarize the key points in the process of gene expression:

The process of gene expression converts information the _____ into the _____.

_____ produces a copy of a gene in the form of _____.

mRNA is used to direct the synthesis of _____ in the process of _____.

Transcription and translation can be broken down into _____, _____, and _____.

_____ gene expression is much more complex than _____ gene expression.

15.9 Mutations: Altered Gene Expression

Recall It

Mutations are heritable changes in genetic material. Mutations can come in many forms and have positive, negative, or no effect on the phenotype of an individual. Point mutations affect a single site in the DNA. Point mutations can result in the addition, deletion, or substitution of one base for another. Base substitutions, silent mutations, nonsense mutations, frameshift mutations, or triplet repeat expansion mutations can arise from changes in base pairs. Chromosomal mutations can also occur, where portions of a chromosome are deleted, duplicated, inverted, or translocated. Many human disorders are the result of both DNA and chromosomal mutations.

Essential Knowledge covered
3.C.1: Changes in genotype can result in changes in phenotype.

Review It

Identify the mutation based on the description provided:

Description	Mutation
An addition of a single base alters the reading frame in the mRNA	
A segment of chromosome is broken in two places, reserved, and put back together	
The deletion of a single base produces a stop codon in the mRNA	
The substitution for a base changes an amino acid in a protein	
A large portion of the short arm of chromosome 5 is deleted	

Use It

Can genetic mutations ever be beneficial? Why or why not?

---

15 Chapter Review

Summarize It

1. Describe the similarities and difference in gene expression between prokaryotes and eukaryotes.

2. Sickle-cell anemia is a disorder that distorts the shape of the red blood cell. The following two sequences show the normal and abnormal protein.

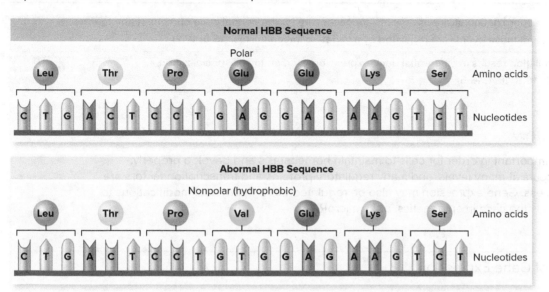

How does the abnormal sequence differ from the normal sequence? How does this change the nature of the protein?

Chapter 16: Control of Gene Expression

Essential Knowledge

2.E.1	Timing and coordination of specific events are necessary for the normal development of an organism, and these events are regulated by a variety of mechanisms. **(16.6, 16.7)**	**Big Idea 2**
3.B.1	Gene regulation results in differential gene expression, leading to cell specialization. **(16.1, 16.2, 16.3, 16.4, 16.5)**	**Big Idea 3**

Chapter Overview

Gene regulation is important in order for cells to maintain homeostasis and develop properly. Gene regulation occurs at many levels, and many regulatory proteins and transcription factors are involved in the process. Gene expression may also be regulated through physical modifications to DNA or histones, as is the case in epigenetics, or by microRNAs.

16.1 Control of Gene Expression

Recall It

Gene control can occur at all levels of gene expression. In transcription, the control of gene expression often occurs through the control of initiation. Regulatory proteins that modulate the ability of RNA polymerase to bind to the promoter may either block transcription or stimulate it. In prokaryotes, gene expression is often driven by changes in the immediate environment. Prokaryotes may need to alter enzymes in response to the type or quantity of nutrients available. The main purpose of gene regulation in eukaryotes is to maintain homeostasis and drive development.

Essential Knowledge covered
3.B.1: Gene regulation results in differential gene expression, leading to cell specialization.

Review It

Define *homeostasis*.

What type of gene regulation is common to both prokaryotes and eukaryotes?

Use It

Why would gene regulation be important to a growing multicellular organism?

16.2 Regulatory Proteins

Recall It

The basis for transcriptional control is the ability for regulatory proteins to bind specific regions of DNA. Regulatory proteins interact with DNA through the major groove of the DNA double helix. DNA-binding domains interact with specific DNA sequences. The regions of a regulatory protein that can bind to the DNA is termed a DNA-binding motif.

Essential Knowledge covered
3.B.1: Gene regulation results in differential gene expression, leading to cell specialization.

Review It

Determine if the following statements regarding regulatory proteins are true or false **(T/F)**:

Regulatory proteins bind to specific DNA sequences.

Regulatory can interact with DNA through the minor groove.

No two regulatory proteins have the same DNA-binding motif.

Regulatory proteins can either block or facilitate transcription.

Use It

Why would it be beneficial for a regulator protein to have protein–DNA-binding sites?

Recall It

In prokaryotes, control of transcription can be either positive or negative. Positive control increases the frequency of transcription, and it is mediated by regulatory proteins called activators. Negative control is mediated by proteins called repressors, which interfere with transcription. Induction occurs when gene expression is turned on in response to a substrate. Repression occurs when gene expression is turned off in the presence of a substrate. An example of induction can be seen in the *lac* operon. The *lac* operon is induced in the presence of lactose. When lactose is present, allolactose binds to the *lac* operon repressor, and the operon is turned on. The *trp* operon is controlled by the *trp* repressor. When tryptophan is present, the enzymes needed to produce tryptophan are turned off. The *trp* operon is repressed when tryptophan, acting as a corepressor, binds to the repressor, altering its conformation such that it can bind to DNA and turn off the operon.

Essential Knowledge covered
3.B.1: Gene regulation results in differential gene expression, leading to cell specialization.

Review It

Define the structures and processes found in prokaryotic regulation.

Structure	Definition
Repression	
Induction	
Operon	
Activators	
Repressors	
Positive Control	
Negative Control	

Use It

The following diagram represents the *trp* operon. Tryptophan is currently missing from the diagram. Using what you learned about prokaryotic regulation in this section, answer the following questions.

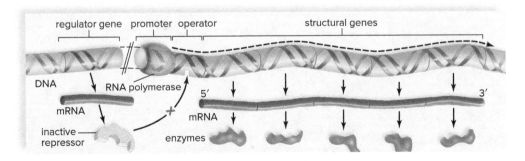

How is the *trp* operon controlled?

Where does tryptophan bind in this pathway if present?

What happens to the operon when tryptophan binds?

16.4 Eukaryotic Regulation

Recall It

The control of transcription in eukaryotes require transcription factors. There are two categories of transcription factors: general and specific. General transcription factors are needed to assemble the transcription apparatus and recruit RNA polymerase II to a promoter. Specific factors stimulate higher rates of transcription in certain cells, tissues, or at certain times through binding to enhancers. Enhancers may be distant from the promoter but can be brought closer by DNA looping. Transcription factors have specific binding sites called promoters and enhancers.

Essential Knowledge covered
3.B.1: Gene regulation results in differential gene expression, leading to cell specialization.

Review It

Identify if the statement applies to a promoter **(P)** or an enhancer **(E)**:

Act in a tissue or time dependent manner.

Mediate the binding of RNA polymerases.

Can act over large distances through DNA looping.

Specific transcription factors bind here.

Use It

The TFIID is a general transcription factor that contains a protein, which recognizes the TATA box sequence in eukaryotes. Where would you expect this sequence to be located? Why?

16.5 Eukaryotic Chromatin Structure

Recall It

Eukaryotic transcription can be regulated through methylation of DNA, modification of histones, or through noncoding DNA. Methylation of DNA bases, often the addition of a methyl group to cytosine, can inactivate genes. Histones can also be methylated, as well as acetylated and phosphorylated. Histone acetylation is often correlated to active sites of transcription. Chromatin-remodeling complexes may also lead to changes in chromatin structure. These complexes contain enzymes that move, reposition, and transfer nucleosomes and ultimately change chromatin structure.

Essential Knowledge covered
3.B.1: Gene regulation results in differential gene expression, leading to cell specialization.

Review It

List three ways chromatin structure can be altered.

Use It

Describe how chromatin-remodeling complexes make DNA more accessible to regulatory proteins.

Recall It

After transcription, small RNAs can alter gene expression through selective mRNA degradation, inhibition of translation, or through altering the structure of chromatin. There are different categories of small RNA, including microRNA and small interfering RNA. Single genes can also be alternatively spliced in response to specific tissues or developmental cues to create multiple mRNA sequences. RNA editing can alter mRNA after transcription. Initiation of translation can also be controlled through translation factors and repressor proteins.

Essential Knowledge covered
2.E.1: Timing and coordination of specific events are necessary for the normal development of an organism, and these events are regulated by a variety of mechanisms.

Review It

Name two categories of small RNA.

How do small RNAs regulate gene expression in eukaryotes?

Use It

The cells of the thyroid and the hypothalamus produce two different hormones: calcitonin and CGRP. These hormones have different physiological purposes. How is it possible that these hormones are produced from the same transcript? What controls which hormone is produced?

16.7 Protein Degradation

Recall It

Protein degradation is a necessary part of a cell lifecycle. There are enzymes called proteases that can degrade protein into amino acids. A second way eukaryotic cells can degrade proteins is through placing molecular markers on proteins called ubiquitin. Proteins with these molecular markers may be then degraded by a cell organelle called the proteasome.

Essential Knowledge covered
2.E.1: Timing and coordination of specific events are necessary for the normal development of an organism, and these events are regulated by a variety of mechanisms.

Review It

Identify the following structures that degrade a protein:

A protein that degrades protein

An organelle which degrades proteins with ubiquitin markers

Use It

Why would systematic protein degradation be important to the longevity of a cell?

16 Chapter Review

Summarize It

E. coli can metabolize lactose through the transcription on of the *lac* operon. Scientists found that in the absence of lactose, the operon is not transcribed. A protein was bound to the operator. The *lac* operon was induced when lactose was added. Using this information, propose an explanation for how the *lac* operon is regulated under differing environmental conditions.

Chapter 17: Biotechnology

Essential Knowledge

3.A.1	DNA, and in some cases RNA, is the primary source of heritable information. (17.1, 17.2, 17.3, 17.4, 17.6, 17.7)	**Big Idea 3**

Chapter Overview

Humans have been manipulating biological systems for thousands of years for environmental, medical, and agricultural purposes. The manipulation and engineering of genetic material makes up the field of biotechnology. Biotechnology has resulted in new forensic tools, as well as new ways to produce medicines, diagnose disease, and manage waste and farm more efficiently.

17.1 Recombinant DNA

Recall It

Restriction endonucleases are enzymes that cleave DNA at specific sites. Scientists use them to produce DNA fragments. DNA fragments can be separated based on size through electrophoresis. DNA has a negative charge, so scientists can separate it using an electric field. DNA is placed in a gel matrix, and an electric current is passed through the gel, causing DNA to migrate to the positive pole. Some DNA fragments are combined to make recombinant molecules- a hybrid made from DNA from two different sources. DNA ligase is used to form phosphodiester bonds in recombinant molecules. The enzyme reverse transcriptase uses RNA as a template to make a DNA molecule. This DNA is called cDNA. DNA libraries store collections of recombinant DNA molecules.

Essential Knowledge covered
3.A.1: DNA, and in some cases RNA, is the primary source of heritable information.

Review It

In your own words, describe the following biotechnology tools:

Biotechnology Tool	Description
Restriction enzyme digest	
Gel electrophoresis	
Recombinant DNA	
cDNA	

Use It

Three distinct bacterial colonies (1, 2, and 3) grew on a plate swabbed with residue on the inside of a dirty coffee mug. The DNA was extracted and analyzed. A highly conserved region of DNA in each bacterial genome was isolated and cut with the same restriction enzymes. The results were read on a gel.

What is the method of DNA separation used to analyze the size of DNA fragments?

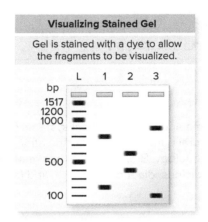

Visualizing Stained Gel

Gel is stained with a dye to allow the fragments to be visualized.

Are the three bacterial colonies the same species? How can you tell?

Which bacterial colony has the largest DNA fragment? How do you know?

Recall It

The polymerase chain reaction (PCR) amplifies a single small DNA fragment thousands of times to produce millions of DNA copies. This involves attaching two short primers flanking the region to be amplified, and running the DNA through cycles of heating and cooling to separate the DNA, anneal the primers, and synthesizes DNA. New DNA is synthesized with additional nucleotides and the thermostable Taq polymerase. PCR is the basis of many sequencing technologies, and without this technology, it would be impossible to sequence genomes. PCR can be coupled with cDNA to amplify specific gene sequences that lack regulatory elements and introns. Quantitation of a specific mRNA is possible through combining reverse transcription to make cDNA with PCR amplification in the presence of fluorescent DNA binding dyes or special fluorescently labeled primers.

Essential Knowledge covered
3.A.1: DNA, and in some cases RNA, is the primary source of heritable information.

Review It

State the three steps of PCR.

List three components needed for PCR.

Use It

What major role does the thermophilic bacterium, *Thermus aquaticus* play in PCR?

17.3   Creating, Correction, and Analyzing Genetic Variation

Recall It

The variation we see in the genetic code between individuals and species allows us to draw greater conclusions about relationships and phenotypes. Forensics uses DNA fingerprinting to identify individuals, often through analyzing short tandem repeats (STRs) that vary in number between individuals in a population. Other tools in biotechnology purposefully create variations in genes to study the change in phenotype, either through using chemical mutagens or in vitro with tools such as PCR or RNAi. Gene editing can remove gene function or introduce new alleles in a chromosome and is

Essential Knowledge covered
3.A.1: DNA, and in some cases RNA, is the primary source of heritable information.

Review It

Describe the difference between genotype and phenotype.

Use It

Someone stole the cookies from the cookie jar, but they made a critical mistake: they left their DNA behind all over an empty glass of milk at the scene of the crime. Taking cookie theft very seriously, the owner had a DNA profile run and tested against the DNA on the three suspects: Who Me, Yes You, and Couldn't Be. With this information in mind, answer the following questions.

There was only a trace amount of DNA recovered from the empty glass. How did scientists increase the amount so they could perform more accurate tests on it? Describe the process in detail.

The scientist working on the case only amplified a portion of the DNA which codes for the human *CookieMnstr* gene. What tool did the scientist use to do this?

Looking at the results, who did it? How can you tell?

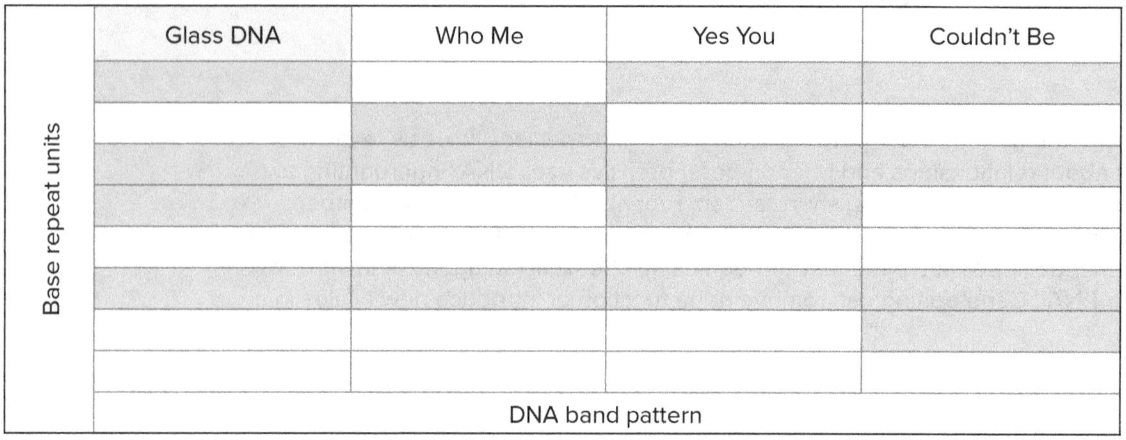
Glass DNA
Who Me
Yes You
Couldn't Be
Base repeat units
DNA band pattern

17.4 Constructing and Using Transgenic Organisms

Recall It

As you may recall from Chapter 14, almost all organisms use the same genetic code. Consequently, genes from one organism can be expressed in other organisms. Organisms that are created and express genes of other species are called transgenic. Humans have bioengineered all sorts of transgenic bacteria, mammals, and plants. Plasmid-based transformations are one technique used to create transgenic bacteria and plants. A gene of interest is inserted into a bacterial plasmid and the altered bacterium is then used to infect another cell.

Essential Knowledge covered

3.A.1: DNA, and in some cases RNA, is the primary source of heritable information.

Review It

Label the plasmid and the gene of interest in this transformed bacterial cell.

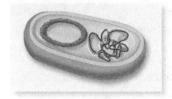

Use It

Create a drawing that shows and explains how a bacterial plasmid can be used to infect a plant cell and turn the plant into a transgenic organism.

17.5 Environmental Applications

Recall It

Biotechnology can be used to protect, repair, and reduce human impacts on the environment. One area of environmental biotechnology is the production of biofuels from plants and algae. Another area includes bioremediation, the use of microorganisms and plants to clean wastewater and/or remove pollutants from the soil or water.

Review It

Biofuels and bioremediation are two examples of biotechnology mentioned in this section. Pick one of the two topics and explain what biological system is manipulated and how it benefits humanity.

Recall It

Biotechnology is leading to many advancements in healthcare. Novel diagnostic tools and treatment techniques allow us to identify and treat diseases and disorders that were once considered untreatable. Genetic engineering techniques, such as the use of recombinant DNA, have revolutionized medicine production. There are now recombinant yeast and bacteria capable of producing medicine, such as human insulin, increasing the potency and purity of pharmaceuticals. Chromosomal alterations involved in disease can be identified through fluorescently labeled DNA probes hybridized to complementary DNA sequences. Gene chips are another tool that can help analyze patterns of gene expression associated with disease. Identification of amplifications of the gene can help design personalized treatments and led to early clinical intervention. Immunoassays, antibodies, stem cell therapy, and bone marrow transplants are some examples of biotechnology used in diagnostic, detection, and treatments of infections and disease.

Essential Knowledge covered
3.A.1: DNA, and in some cases RNA, is the primary source of heritable information.

Review It

Many advancements in biotechnology require a thorough understanding of molecular biology. Define the molecular biology vocabulary below:

Vocabulary	Definition
Exon	
Polypeptide	
Promoter	
Denatured	
mRNA	
Recombinant DNA	

Use It

In humans, two insulin chains, chain A and chain B, are required to produce active insulin. The figure below shows the first few steps in the production of human insulin in genetically engineered *E. coli* bacteria.

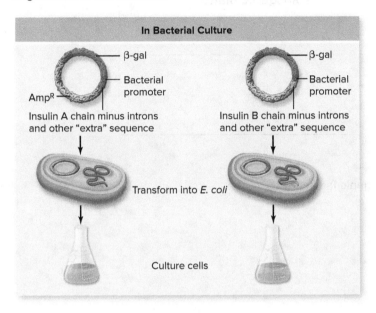

In Bacterial Culture

β-gal

Bacterial promoter

Amp^R

Insulin A chain minus introns and other "extra" sequence

β-gal

Bacterial promoter

Insulin B chain minus introns and other "extra" sequence

Transform into *E. coli*

Culture cells

What are the next steps in this genetically engineered insulin production?

How is this process different from the production of insulin in humans?

17.7 Agricultural Applications

Recall It

The field of agriculture has been greatly impacted by biotechnology. Transgenic crops have been created that are herbicide-resistant, insect resistant, and/or contain more vitamins (such as golden rice). While such crops reduce the need for tilling, the associated fossil fuel consumption, insecticide, or may provide a more nutritious diet, transgenic crops raise a number of social issues. There are concerns about allergies, safety of human consumption, and how transgenic plants will change biodiversity.

Essential Knowledge covered
3.A.1: DNA, and in some cases RNA, is the primary source of heritable information.

Review It

Identify the type of transgenic plant based on the description provided:

Description	Transgenic plant
Produces a toxin from the bacterium *Bacillus thurigniensis*	
Carries resistance to glyphosphate	
Expresses β-carotene	

Use It

In what way do transgenic crops pose a problem for organic famers?

17 Chapter Review

Summarize It

A home gardener living near a commercial sunflower farm suddenly found that the weeds in his garden were becoming unmanageable. While he once could remove these weed species with one herbicide application, the weeds now seemed to be resistant. Provide a reasonable explanation for the change in the weeds' physiology.

Chapter 18: Genomics

Extending Knowledge

In Chapter 1, you learned that the theory of evolution is the unifying theory of biology, and is supported by multiple lines of evidence. Genomics, detailed in Ch. 18, is one of these lines of evidence.

Big Idea 1

Big Idea 3

Chapter Overview

Genome sequencing and assembly produces vast amounts of data which are stored in large databases. Genomes contain both coding and non-coding sequences. Traditionally, scientists have been more interested in the coding regions, but the importance of noncoding regions has become more apparent. Genomics has provided incredible advancements for humanity but carries with it notable ethical and moral concerns.

Review It

Describe the classes of DNA sequences found in the human genome.

Class	Description
Noncoding RNA	
Transposable elements	
Structural DNA	
Pseudogenes	
Genes	

Determine what type of genomics you would use to answer the following question: comparative genomics **(C)** or functional genomics **(F)**.

What is the structure of protein that causes cystic fibrosis?

How closely related are ducks and geese?

How does gene expression change in an embryo during development?

Do green algae and red algae perform photosynthesis in the same way?

Describe a way in which genomics benefits humanity.

Summarize It

Currently, the United States' National Institute of Health (NIH) believes that editing the DNA of a human embryo is morally wrong. Do you take the same stance on the idea on genetically engineering humans? Provide one reason that genetically engineering a human embryo may be considered "right" or "wrong."

Chapter 19: Cellular Mechanisms of Development

Extending Knowledge

In Chapter 14, you learned that DNA, and sometimes RNA, is the primary molecule containing genetic information. Scientists can use their knowledge of the structure and function of DNA in genetic engineering, which allows them to better understand the processes of development covered in Chapter 19.

Big Idea 3

Chapter Overview

Through your study of biology thus far, you have learned that DNA carries all the information an organism needs for growth and development. Gene expression is controlled through diverse and interacting mechanisms in many organisms. Multicellular organisms have incredibly complicated gene patterns and developmental pathways. While many of the cellular changes during development are irreversible, the cloning of Dolly the sheep showed us that the nucleus of an adult cell can be reprogrammed to be totipotent. Because of the complexities of gene expression, reproductive cloning has a low success rate, and clones often develop age-associated diseases.

Review It

The following molecular biology terms are important to know when considering the process of reproductive cloning. Fill in the definition for the following terms:

Vocabulary	Definition
Zygote	
Nucleus	
Cytoplasm	
Genomic imprinting	

Summarize It

In what way did the existence of Dolly the sheep shows that a differentiated adult cell can be used to drive all of development? What aspects of Dolly's life indicated the complexities of gene expression?

Chapter 20: Genes Within Populations

Essential Knowledge

1.A.1	Natural selection is a major mechanism of evolution. **(20.1, 20.2, 20.3, 20.4, 20.5, 20.7, 20.9)**	**Big Idea 1**
1.A.2	Natural selection acts on phenotypic variations in populations. **(20.5)**	**Big Idea 1**
1.A.3	Evolutionary change is also driven by random processes. **(20.3, 20.8)**	**Big Idea 1**
3.C.1	Changes in genotype can result in changes in phenotype. **(20.3, 20.6)**	**Big Idea 3**

Chapter Overview

In order for evolution to occur within a population, the population must contain genetic variation. Natural selection, a major mechanism of evolution, acts on this genetic variation. This chapter describes how changes in allele frequency affect evolution, agents that drive evolution, and how evolution is studied.

20.1 Genetic Variation and Evolution

Recall It

Genetic variation is the number of different alleles found within a population. Genetic variation is necessary for evolution. Natural selection often drives evolution; when some individuals in a population produce more offspring than others, evolutionary change can occur. Other processes drive evolution as well, such mutations in alleles. DNA testing shows that natural populations generally have substantial variation.

Essential Knowledge covered
1.A.1: Natural selection is a major mechanism of evolution.

Review It

In your own words, what is evolution?

Define *population genetics.*

Use It

The groove snail, *Cepaea nemoralis,* is found all over Western Europe. Some snails have no bands, and some are have one or more bands. The color of the shell can vary too; snails can be yellow, brown, pink, or white. Provide an explanation for the color and stripe variation in the groove snails. Why might a darker colored groove snail be found in the woods and a lighter snail in a garden bed?

Recall It

The Hardy–Weinberg principle predicts the genotypic frequencies in a population. When the observed genotype frequencies match the prediction from calculated frequencies, a population is said to be in equilibrium. Equilibrium occurs only when evolutionary processes are not acting to shift the distribution of alleles or genotypes in the population. If genotype frequencies are not in Hardy–Weinberg equilibrium, then evolutionary processes must be at work.

Essential Knowledge covered
1.A.1: Natural selection is a major mechanism of evolution.

Review It

List the five conditions that must be met in order for a population to be in Hardy–Weinberg equilibrium.

Provide the algebraic formula for Hardy–Weinberg equilibrium. What does each piece of the equation represent?

Use It

Estimate the genotype frequencies of a population at Hardy–Weinberg equilibrium given the following allele frequencies: 0.75 W, 0.25 w

Recall It

There are five agents of evolutionary change: (1) mutation, (2) gene flow, (3) nonrandom mating, (4) genetic drift, and (5) selection. These agents allow populations to evolve over time. Mutations are the ultimate source of genetic variation, but because mutation rates are low, mutation usually is not responsible for deviations from Hardy–Weinberg equilibrium. Gene flow is the migration of new alleles into a population. Depending on the source population, gene flow can either increase genetic variation or homogenize allele frequencies. Nonrandom mating shifts genotype frequencies. In assortative mating, similar individuals tend to mate increasing the portion of homozygotes in the next generation. During disassortative mating, phenotypically different individuals mate and increase the frequency of heterozygotes. Genetic drift refers to random shifts in allele frequency in small populations. Selection favors some genotypes over others and produces adaptive evolutionary change. In order for evolution by natural selection to occur, three conditions must be met: (1) genetic variation must exist, (2) genetic variation must result in differential reproductive success, and (3) genetic variation must be heritable.

Essential Knowledge covered
1.A.1: Natural selection is a major mechanism of evolution.
1.A.3: Evolutionary change is also drive by random processes.
3.C.1: Changes in genotype can result in changes in phenotype.

Review It

Based on the examples below, identify the agent of evolutionary change:

Example	Agent of Evolutionary Change
Pollen is carried a great distance by a bee, introducing new alleles into a clover patch.	
A small population of goats are stranded on an island and become distinctly smaller in size then goats on the mainland.	
Dandelions self-fertilize and create a population of homozygous individuals.	
A population of lizards living on dark rock formations are dark in color	
A chemical changes the nucleotide sequence in a developing gamete and produces a new trait.	

Use It

In this section, you learned about different types of genetic drift. In the table below, define the bottleneck effect with the founder effect, and describe commonalities between them.

Bottleneck Effect	Commonalities	Founder Effect

A population of bacteria in a river downstream from effluent from a hospital were sampled and found to be different then the population upstream of the effluent. The population downstream had a genotype that provided the bacteria a greater resistance to antibiotics. Describe the agent of evolutionary change that may be at work on the downstream population of bacteria.

20.4 Quantifying Natural Selection

Recall It

Natural selection can be quantified through fitness and relative fitness. Fitness is the reproductive success of an individual , also measured as the number of surviving offspring left in the next generation. Relative fitness refers to the success of one phenotype relative to others in a population. A phenotype with greater fitness usually increases in frequency. Likewise, the genotype with highest relative fitness increases in frequency in the next generation. Fitness may consist of many components, including survival, mating success, and number of offspring or reproductive success. Reproductive success is determined by how long an individual survives how often it mates, and how many offspring it has per reproductive event.

Essential Knowledge covered
1.A.1: Natural selection is a major mechanism of evolution.

Review It

Define *fitness*.

List three factors that may determine reproductive success.

Use It

Using the following dataset, what is the difference in fitness between the populations? Which population has greater fitness, and why?

	Population A	**Population B**
Phenotype	Spikey	Smooth
Number of offspring	20	4

If the difference in the above population had a genetic basis, and the fitness of the two populations remained the same, which allele would eventually be removed?

20.5 Natural Selection's Role in Maintaining Variation

Recall It

While natural selection may change populations through favoring one allele over others, it can also help maintain population variation as well. Frequency-dependent selection may favor either rare or common phenotypes. In negative frequency-dependent selection, a rare phenotype is favored and maintained within a population. In positive frequency-dependent selection, the common phenotype is favored and leads to decreased variation. The favored phenotype changes as the environment changes, leading to oscillating selection. When environmental change is cyclical, selection favor first one phenotype, then another. In this manner, variation is maintained. In some cases, heterozygotes may exhibit greater fitness than homozygotes, and alleles that are deleterious in the homozygous state are retained. This is known as the heterozygote advantage.

Essential Knowledge covered
1.A.1: Natural selection is a major mechanism of evolution.
1.A.2: Natural selection acts on phenotypic variations In populations.

Review It

Explain how the following processes affect genetic variation within a population:

Process	Effect
Negative frequency-dependent selection	
Positive frequency-dependent selection	
Oscillating selection	
Heterozygote advantage	

Use It

Sickle cell anemia is a hereditary disease that affects the structure of hemoglobin. The homozygous form of the disease is usually fatal, and individuals with both homozygous alleles rarely make it to reproductive age. The disease is common in central Africa, where malaria is also present. Provide a reasonable explanation of why the homozygous allele for this disease has not been removed from the population through natural selection.

20.6 Selection Acting on Traits Affected by Multiple Genes

Recall It

As you learned in Chapter 12, multiple genes may control one trait, and such traits exhibit a continuous or normal distribution. Selection can alter the normal distribution of a trait within a population. Disruptive selection removes intermediary traits, creating a population with a bimodal trait distribution. Directional selection eliminates phenotypes at one end of a range, shifting the mean value of the population toward the favored end of the distribution. Finally, stabilizing selection favors individuals with intermediate phenotypes, eliminates extremes and increases the frequency of an intermediate type.

Essential Knowledge covered
3.C.1: Changes in genotype can result in changes in phenotype.

Review It

Using your textbook as a reference, determine the types of selection illustrated below:

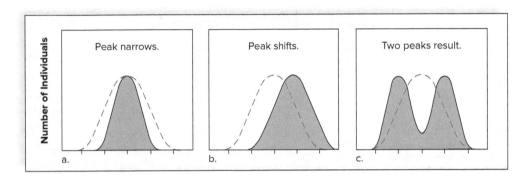

Use It

Stabilizing selection favors individuals with intermediate phenotypes. Stabilizing selection is often seen in ducks; duck eggs of intermediate weight often have the highest hatching success. Why do you think this is?

20.7 Experimental Studies on Natural Selection

Recall It

The hypothesis that natural selection leads to evolutionary change can be tested experimentally. An example presented to you in this section described a laboratory and field experiment on the color variation and reproductive success of guppies found in different environments. Guppies in both natural and laboratory populations subject to different predators were shown to undergo color change over generations.

Essential Knowledge covered
1.A.1: Natural selection is a major mechanism of evolution.

Review It

Fill in the scientific process that the scientists in this section followed to answer the question "Does the presence of predators affect the evolution of guppy color?"

Hypothesis:	
Experiment:	

Use It

Review the guppy color experiment on pp. 414–415 of your textbook. The results of the study are presented in the graph below. After you have completed your review, answer the questions below:

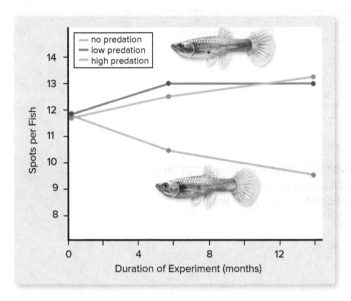

Describe the results of this experiment.

Why do you think colorful spots are beneficial in subsequent generations without predators?

20.8 Interactions Among Evolutionary Forces

Recall It

Different evolutionary processes can work together or against each other. While it rarely occurs, a high rate of mutation could oppose natural selection. Gene flow may impede adaptation due to influx of alleles with low fitness in a population's environment.

Essential Knowledge covered
1.A.3: Evolutionary change is also driven by random processes.

Review It

Provide an example of how mutations and gene flow can be a constructive or constricting force to a population.

Evolutionary Process	Constructive	Constraining
Mutation		
Gene Flow		

Use It

Some grasses have an allele for copper tolerance. Grasses with this allele can grow in contaminated soils, but they grow more slowly. Describe how copper tolerance in grass on and near old mine sites can spread undesirable alleles into neighboring grass populations through gene flow.

20.9 The Limits of Selection

Recall It

Pleiotropy and epistasis place restraints on the effects of natural selection. Pleiotropic genes are genes that have multiple effects on a phenotype. The change in one allele may have a beneficial effect in one area but be hindered by a secondary effect. There can be no evolution if there is little genetic variation or if phenotypic variation does not have a genetic basis. Epitasis also places a limit on selection as there may be a selective advantage of one allele from one genotype to another.

Essential Knowledge covered
1.A.1: Natural selection is a major mechanism of evolution.

Review It

Review the vocabulary words presented in Chapter 12 needed to understand the limits on selection:

Pleiotropy:

Epistasis:

Use It

Identify the force limiting natural selection:

Example	Limiting Force
Thoroughbred horses have not improved in speed in 50 years following several hundred years of intense selection.	
An allele that increase the size of a tomato also decreases the flavor of the fruit.	
The fitness of one allele depends on the genotype of the second gene.	

20 Chapter Review

Summarize It

Use the following scenario to answer the following questions:

Suppose the observed frequencies of the genotypes PP, Pp, and pp in a population of lilies were 0.5, 0.25, and 0.25 respectively. Lilies with the dominant allele had pink flowers.

   a. What percent of the flowers are pink?

   b. Is the population in Hardy–Weinberg equilibrium?

   c. What might cause this population not to be in Hardy–Weinberg equilibrium?

Chapter 21: The Evidence for Evolution

Essential Knowledge

1.A.2	Natural selection acts on phenotypic variations in populations. **(21.2, 21.3)**	Big Idea 1
1.A.4	Biological evolution is supported by scientific evidence from many disciplines, including mathematics. **(21.4, 21.5)**	Big Idea 1
1.C.1	Speciation and extinction have occurred throughout the Earth's history. **(21.6)**	Big Idea 1
1.C.3	Populations of organisms continue to evolve. **(21.1)**	Big Idea 1

Chapter Overview

Evolution is supported though evidence from many fields, including anatomy, molecular biology, biogeography, genetics, and geology. Darwin's theory is sustained through evidence that natural selection can produce evolutionary change. This is reinforced through changes seen in the fossil record.

21.1   The Beaks of Darwin's Finches: Evidence of Natural Selection

Recall It

Darwin's finches are one of the greatest examples of evidence for natural selection. When Darwin traveled to the Galápagos Islands in 1893, he noticed that the different species of finches had very different beak sizes and shapes. Darwin noted that beak variation related to what types of foods the finches ate and how they gathered or obtained food. Darwin hypothesized that natural selection had acted on beak morphology, favoring larger-beaked birds during extended droughts and smaller-beaked birds during long periods of heavy rains. Modern research has verified Darwin's selection hypothesis. The variation in beak size in finches is heritable, allowing evolutionary change to occur in the beak sizes of subsequent generations.

Essential Knowledge covered
1.C.3: Populations of organisms continue to evolve.

Review It

Review the three conditions that must be met in order for evolution to occur by natural selection:

Use It

Well over 50 years after Darwin's travels to the Galápagos Islands, researchers returned test his hypothesis that beak size evolved to adapt to different food sources. Use the graph to describe what the researchers studied and what they found:

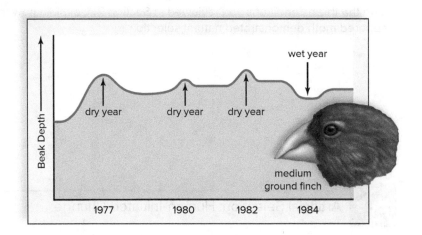

21.2   Peppered Moths and Industrial Melanism: More Evidence of Selection

Recall It

The peppered moth remains a classic example of natural selection. Peppered moths come in two morphs: dark and light. The dark-colored morph became dominant in polluted areas where soot built up on tree trunks. In unpolluted areas, the light-colored form remained predominant. When the polluted conditions reversed, the frequency of light-colored moths in polluted areas increased. It was suggested that predation by birds was the cause; light-colored moths were easier to spot when they landed dark trunks; whereas, dark-colored moths were able to blend in with the soot covered tree trunks. Regardless, the observation the industrial melanism seen in peppered months supports the idea of natural selection; when the environment changes, a previously unfavorable trait may now provide greater reproductive success.

Essential Knowledge covered
1.A.2: Natural selection acts on phenotypic variations In populations.

Review It

Determine if the following statements are true or false **(T/F)** regarding industrial melanism in peppered moths:

White moths gained a survival advantage in polluted areas around 1850 in England.

The peppered moth was made almost extinct by industrial pollution.

All-black peppered moths are the result of a single mutation.

Dark moths became common in areas where soot had darkened tree trunks.

Use It

Using the three conditions you reviewed in Section 21.1, explain how industrial melanism in the peppered moth demonstrated natural selection.

21.3 Artificial Selection: Human-Initiated Change

Recall It

Humans impose artificial selection on other organisms to produce desired change in populations. Extensive laboratory experiments have shown that evolutionary change can occur in controlled populations. Humans have long selected for desirable traits in crops and livestock. Many domesticated breeds and crop plants are now substantially different from their wild ancestors. Artificial selection can rapidly create substantial change over short periods.

Essential Knowledge covered
1.A.2: Natural selection acts on phenotypic variations in populations.

Review It

Determine if the following scenarios demonstrate artificial **(A)** or natural selection **(N)**:

After repeated exposure to penicillin, a population of antibiotic-resistant bacteria develops in a hospital.

Russian researchers breed silver foxes for desired traits, such as friendliness, for several generations.

Gray tree frogs survive longer and reproduce more often in forests that contain many lichens than green tree frogs.

Farmers select seeds from their largest ears of corn to plant for next year's crop.

Use It

Compare and contrast artificial and natural selection, providing examples of each.

Recall It

The fossil record provides an incredible map of the evolution of life. New fossil species are constantly being discovered. Slowly, a detailed understanding of major evolutionary transitions is emerging. The date of fossils in rock can be estimated by measuring the extent of radioactive decay based on half-lives of known isotopes. The fossil record shows that, over the course of 50 million years, evolution has not been constant and uniform. During some periods of evolutionary history, change has been rapid, but sometimes transitions have been very slow.

Essential Knowledge covered
1.A.4: Biological evolution is supported by scientific evidence from many disciplines, including mathematics.

Review It

Define *fossil.*

Use It

The following illustration shows ancestors of the modern-day horse, placed in chronological order of their existence. Fossils of the toes and molar teeth are shown above each species. Do you spot any trends in the evolution of the horse? If *Nannippus* had never been discovered by geologists, how might this change your answer?

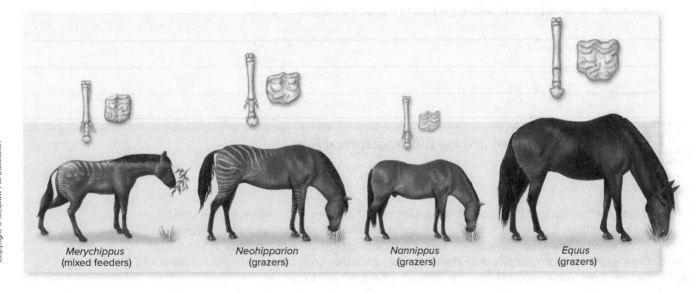

Merychippus
(mixed feeders)

Neohipparion
(grazers)

Nannippus
(grazers)

Equus
(grazers)

Recall It

Evidence for evolution can be observed through the study of anatomy. Natural selection has modified the same structure in different species to produce homologous structures. Homologous structures have different appearances and functions but were derived from the same common ancestral body part. Strong anatomical evidence of evolution can often be seen during early embryonic development in species that appear vastly distinct during adulthood. As natural selection can only work on the variation present in a population, some structures are imperfectly suited to their use. Vestigial structures have no apparent function but resemble ancestral structures. Pseudogenes are traces of previously functioning genes.

Essential Knowledge covered
1.A.4: Biological evolution is supported by scientific evidence from many disciplines, including mathematics.

Review It

Provide an example of for each type of structure:

Structure	Example
Homologous	
Vestigial	

Use It

The modern day baleen whale has vestigial pelvic bones. Draw, label, and explain what you think its ancestor may have looked like before the bones became vestigial.

Recall It

Biogeography is the study of the geographic distribution of species. Frequently, species that live in distinctly different geographical areas evolve in parallel to exhibit similar phenotypes. This phenomenon is known as convergent evolution. Convergent evolution occurs in species or populations exposed to similar selective pressures. Species that colonize islands often evolve into more species great morphological diversity through adaptive radiation.

Essential Knowledge covered
1.C.1: Speciation and extinction have occurred throughout the Earth's history.

Review It

Would you expect a mammal living in water and a mammal living in a desert to have similar features? Why or why not?

Use It

Barracuda are bony fish and dolphins are mammals. Both are predators with long streamline bodies and sharp teeth. How did these distantly related species evolve similar traits?

21.7 Darwin's Critics

Recall It

Biologist now almost universally accept Darwin's theory of evolution by natural selection. The present and historic criticisms made of Darwin's theory mostly stem from a lack of understanding of how scientific theories are formed, the theory's actual content, or the time spans involved in evolution.

Review It

Select one of the seven arguments against Darwin's theory of evolution listed in section 21.7 and provide a scientific rebuttal to the argument.

21 **Chapter Review**

Summarize It

The lesser anteater in South America and the numbat of Western Australia have similar features and similar diets. What steps would you take to determine if these organisms are closely related to one another or are similar because of convergent evolution?

Lesser anteater

Numbat

Chapter 22: The Origin of Species

Essential Knowledge

1.A.1	Natural selection is a major mechanism of evolution. **(22.3)**	**Big Idea 1**
1.C.1	Speciation and extinction have occurred throughout the Earth's history. **(22.5, 22.6, 22.7)**	**Big Idea 1**
1.C.2	Speciation may occur when two populations become reproductively isolated from each other. **(22.1, 22.2, 22.3, 22.4)**	**Big Idea 1**
1.C.3	Populations of organisms continue to evolve. **(22.5)**	**Big Idea 1**
2.E.2	Timing and coordination of physiological events are regulated by multiple mechanisms. **(22.1)**	**Big Idea 2**

Chapter Overview

The biological species concept says a species is reproductively isolated from all other species. Speciation may occur during either prezygotic or postzygotic periods, through many different mechanisms. Adaptive radiation may lead to rapid diversification when species find themselves in a new or changed environment with many resources and little competition.

22.1 The Nature of Species and the Biological Species Concept

Recall It

Sympatric species occupy the same geographical region but are different in phenotype, niche, or behavior. When populations of a single species are separated geographically, they may develop distinct phenotypes or behaviors. There are several types of reproductive isolation mechanism. Prezygotic isolating mechanism include: (1) ecological isolation, (2) behavioral isolation, (3) temporal isolation, (4) mechanical isolation, and (5) prevention of gamete fusion. Postzygotic isolation mechanisms occur when hybrid embryos are unable to develop properly, when hybrid adults cannot survive to reproductive age, or when hybrid adults are sterile.

Essential Knowledge covered
1.C.2: Speciation may occur when two populations become reproductively isolated from each other.
2.E.2: Timing and coordination of physiological events are regulated by multiple mechanisms.

Review It

Provide a description for the following mechanisms of reproductive isolation:

Reproductive Isolation Mechanism	Description
Ecological isolation	
Behavioral isolation	
Temporal isolation	
Mechanical isolation	
Hybrid infertility	

Use It

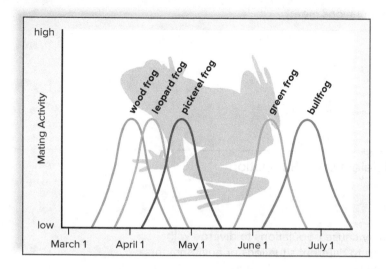

The figure above illustrates the breeding times of different species of frogs in the genus *Rana*. Describe what reproductive isolation mechanism likely led to their speciation.

22.2 Natural Selection and Reproductive Isolation

Recall It

Populations may evolve into two different species following complete reproductive isolation. In addition, natural selection can reinforce isolating mechanisms in populations that are not yet completely isolated. Genetic exchange between the populations can lead to homogenization and prevent speciation from occurring.

Essential Knowledge covered
1.C.2: Speciation may occur when two populations become reproductively isolated from each other.

Review It

Review and define the following evolutionary terms:

Evolutionary Process	Definition
Natural selection	
Reproductive isolation	
Gene flow	

Use It

What might happen when two incompletely isolated populations meet?

22.3 The Role of Genetic Drift and Natural Selection in Speciation

Recall It

Randomly generated genetic change or genetic drift may cause populations to diverge with differences that cause reproductive isolation. Adaptation to different situations or environments acted upon by natural selection may also lead to speciation.

Essential Knowledge covered
1.A.1: Natural selection is a major mechanism of evolution.
1.C.2: Speciation may occur when two populations become reproductively isolated from each other.

Review It

List two major types of genetic drift.

Use It

Describe the difference in process of speciation of a population of lizards that were (A) carried accidently by boat from a larger population to a small island, and developed different mating behaviors than the original population over time, and (B) developed different mating behaviors in open habitats versus wooded habitats.

22.4 The Geography of Speciation

Recall It

Allopatric speciation takes place when populations are geographically isolated. Allopatric populations may evolve into separate species because there no gene flow between them. Sympatric speciation occurs without geographic separation. One type of instantaneous sympatric speciation is polyploidy. Disruptive selection is an extremely rare form of sympatric speciation.

Essential Knowledge covered
1.C.2: Speciation may occur when two populations become reproductively isolated from each other.

Review It

Which type of speciation is the primary means of speciation?

Use It

Allopolyploidy is a very common type of polyploidy speciation that often occurs in plants. Describe how this event occurs and how it might lead to a new species of plant.

22.5 Adaptive Radiation and Biological Diversity

Recall It

When a species finds itself in a new or suddenly changed environment with many resources and few competing species, adaptive radiation may occur. Key innovations are newly evolved traits that allow species to use previously inaccessible resources. Hawaiian *Drosophila*, Darwin's finch species, and Victoria cichlid fishes are three cases that highlight rapid speciation under conditions of isolation.

Essential Knowledge covered
1.C.1: Speciation and extinction have occurred throughout the Earth's history.
1.C.3: Populations of organisms continue to evolve.

Review It

List two conditions that are common in areas where adaptive radiation is known to occur.

Use It

Describe the most likely cause for the speciation of Darwin's finches.

Recall It

Gradualism is the accumulation of small changes and is, historically, how scientists viewed the occurrence of speciation. The contrasting view is punctuated equilibrium, or long periods of stasis followed by relatively rapid change, which has also become a hypothesis for how speciation occurs. Today scientists generally agree that evolutionary change occurs on a continuum, with gradualism and punctuated change being the extremes.

Essential Knowledge covered
1.C.1: Speciation and extinction have occurred throughout the Earth's history.

Review It

Determine if the following statements apply to either gradualism **(G)** or the punctuated equilibrium hypothesis **(P)**:

Speciation can occur gradually with little changes accumulating.

Contains long periods of stasis.

Change happens rapidly.

Use It

Do all species evolve at the same rate?

22.7 Speciation and Extinction Through Time

Recall It

The number of species has increased through time. There have been five mass extinctions in the history of life on Earth, and a sixth one – caused by humans – is currently under way.

Essential Knowledge covered
1.C.1: Speciation and extinction have occurred throughout the Earth's history.

Review It

Define mass extinction.

Use It

Describe the mass extinction that is currently occurring. How many species to scientists expect will become extinct? What do scientists predict for species recovery?

22 Chapter Review

Summarize It

1. The first ancestral cichlid is thought to have entered Lake Victoria 200,000 years ago. In the intervening years, the lake underwent several dramatic level changes: it flooded, dried up into many different pools, and then flooded into one big lake again. The four cichlid fish above are a representative of the many different cichlid species found in Lake Victoria today. Using what you know about speciation, propose an explanation for the speciation of the cichlids of Lake Victoria.

Chapter 23: Systematics, Phylogenies, and Comparative Biology

Essential Knowledge

1.B.2	Phylogenetic trees and cladograms are graphical representations (models) of evolutionary history that can be tested. **(23.1, 23.2, 23.3, 23.4, 23.5)**	**Big Idea 1**

Chapter Overview

Systematics is the study of evolutionary relationships. Evolutionary trees or phylogenies show how species are related using derived characters. Phylogenetic analyses can often help explain species diversification.

23.1 Systematics

Recall It

Systematics is the study of evolutionary relationships. Evolutionary relationships can be depicted using branching evolutionary trees or phylogenies. As the rate of evolution can vary among species, similarities alone may not accurately predict evolutionary relationships. Closely related species can be dissimilar in phenotypic characteristics, and convergent evolution may result in distantly related species with phenotypic similarities.

Essential Knowledge covered
1.B.2: Phylogenetic trees and cladograms are graphical representations (models) of evolutionary history that can be tested.

Review It

Use the following cladogram to answer questions regarding the phylogeny of apes:

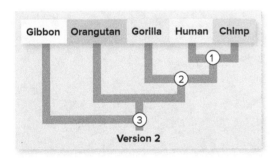

Which ape is the closest living relative to the human?

Are humans more closely related to gibbons or gorillas?

What does Node 3 represent?

Use It

Provide two reasons two phenotypically similar species may not be closely related.

23.2 Cladistics

Recall It

Cladograms are depictions of a hypothesis of evolutionary relationships. Cladograms are constructed using only shared derived characters, such as: morphology, physiology, behavior, and DNA or protein sequences. Species that share a common ancestor are said to belong to a clade. Cladograms can only provide the order of evolutionary branching events, but not the timing involved in the separation between species.

Essential Knowledge covered
1.B.2: Phylogenetic trees and cladograms are graphical representations (models) of evolutionary history that can be tested.

Review It

What is the first step in cladistics analysis?

Use It

Site	DNA Sequence									
	1	2	3	4	5	6	7	8	9	10
Species A	G	C	A	T	A	G	G	C	G	T
Species B	A	C	A	G	C	C	G	C	A	T
Species C	G	C	A	T	A	G	G	T	G	T
Species D	A	C	A	T	C	G	G	T	G	G
Outgroup	A	T	A	T	C	C	G	T	A	T

Using the outgroup DNA sequence as a reference create a cladogram showing the relationships between these species of trees.

Recall It

Classification is the method traditionally used for placing species into taxonomic hierarchy: genus, family, order, class, phylum, kingdom, and domain. Ever-growing phylogenic advancements often reorder previously classified organisms, so that traditional classifications sometimes do not reflect evolutionary relationships. The phylogenetic species concept (PSC) defines a species as any population with one or more derived characters separated from another population, whereas the biological species concept focuses on reproductive isolation. The drawbacks of the PSC are that it can be used to subdivide groups into impractical distinctions, and that the PSC definition of a group may not always apply as selection proceeds.

Essential Knowledge covered
1.B.2: Phylogenetic trees and cladograms are graphical representations (models) of evolutionary history that can be tested.

Review It

Determine if the following statements apply to either the biological species concept **(B)** or the phylogenetic species concept **(P)**:

Focuses on reproductive isolation.

Can be applied to both sexual and asexual species

Focuses on shared derived characters.

Can be applied to allotropic populations.

Use It

Which species concept when applied to an ecosystem would identify more populations as separate species and why?

Recall It

Homologous structures or traits have been derived from the same common ancestor. Homologous traits show that complex characters evolve through a sequence of evolutionary changes, including stages of transition. Homologous traits may have begun as an adaptation to a selective pressure different from the one for which the feature is currently adapted. Different evolutionary scenarios can be distinguished by phylogenetic analysis. Scientists can establish the minimum number of times a trait evolved, and infer the direction and timing of trait evolution as well as the cause of diversification. Phylogenetics helps explain why some clades have a greater number of species or species richness than others.

Essential Knowledge covered
1.B.2: Phylogenetic trees and cladograms are graphical representations (models) of evolutionary history that can be tested.

Review It

Compare and contrast homologous traits with homoplastic traits.

In one sentence, describe the way that complex characters evolve.

Use It

Describe how the phylogenetic analysis of beetle and angiosperm speciation helps explain why 80% of all insects are beetles.

23.5 Phylogenetics and Disease Evolution

Recall It

Phylogenetics can be used to examine relationships among species but is also used to identify the transmission of disease. Phylogenetic methods have indicated that HIV is related to a simian immunodeficiency virus (SIV). It is clear that HIV has descended from SIV, and that independent transfers from simians to humans have occurred several times. Even though HIV evolves rapidly, phylogenetic analysis can trace the origin of a current strain to a specific source of infection.

Essential Knowledge covered
1.B.2: Phylogenetic trees and cladograms are graphical representations (models) of evolutionary history that can be tested.

Review It

List two uses for phylogenetics.

Use It

Describe how phylogenies can be used to track the evolution of AIDS among individuals.

Summarize It

Construct a cladogram, using the chart of the physical traits in the organisms listed below.

Trait	Kangaroo	Human	Koala	Cat	Whales
Placental mammal		X		X	X
Fins					X
Four limbs	X	X	X	X	
Canine teeth		X	X	X	
External tail	X			X	X

Chapter 24: Genome Evolution

Essential Knowledge

1.A.4	Biological evolution is supported by scientific evidence from many disciplines, including mathematics. **(24.1)**	**Big Idea 1**
1.B.1	Organisms share many conserved core processes and features that evolved and are widely distributed among organisms today. **(24.4)**	**Big Idea 1**
3.C.1	Changes in genotype can result in changes in phenotype. **(24.2)**	**Big Idea 3**
4.C.1	Variation in molecular units provides cells with a wider range of functions. **(24.3)**	**Big Idea 4**

Chapter Overview

Studying the structure of genomes provides many clues to evolution. Evolutionary relationships between different species can be determined through genome comparisons. Comparative genomics can also lead to agricultural and medical improvements.

24.1 Comparative Genomics

Recall It

Comparing the genomes of different species allow scientists to study when and how different species diverged. Genomes evolve at different rates but many genes remain conserved. Species that are distantly related to one another may still contain similar genes that were found in a common ancestor. Complex eukaryotic species evolve over millions of years. Viral, bacterial, and even insect genomes evolve more rapidly than mammalian genomes. Plant genomes often evolve faster than animal genomes, possibly as the result of massive genome remodeling from extensive transposition of mobile elements. Plant, fungal, and animal genomes contain both unique and shared genes. It appears about one-third of plant genes are unique to plants. Of the remaining plant genes, many are also found in animal and fungal genomes and are required for metabolism and gene expression.

Essential Knowledge covered
1.A.4: Biological evolution is supported by scientific evidence from many disciplines, including mathematics.

Review It

Define genome.

Order the following genomes from those that are more likely to evolve rapidly from shortest (in a matter of days) to longest (over millions of years): rice, *E.coli,* pufferfish, HIV.

Use It

Using what you learned about comparative genomics in this section, create an explanation of why scientists use mice in human pharmaceuticals testing.

Recall It

As described in Chapter 22, polyploidy can give rise to new species. Genome analysis revealed that polyploidy has occured frequently in flowering plant evolution. Errors in meiosis can cause autopolyploidy, leading to a duplicated genome; allopolyploidy is the result of hybridization between species. Following polyploidy, there is often a rapid loss of genes and/or a rearrangement of chromosomes. Polyploidy can also alter gene expression through methylation of DNA and/or through mobilizing small genetic elements called transposons. Transposons become highly active after polyploidization; their insertions into new positions may lead to new phenotypes. Polyploidy alone does not account for variation in genome size. Genome size is most often inflated due to the presence of introns and non-protein-coding sequences. Genome size does not correlate with the number of genes. Some species have very little noncoding DNA, and others have extensive amounts of noncoding DNA.

Essential Knowledge covered
3.C.1: Changes in genotype can result in changes in phenotype.

Review It

If an organism has a giant genome, does this means it has many genes? Why or why not?

Use It

What is a transposon, and how might it change the phenotype of an organism?

24.3 Evolution within Genomes

Recall It

Aneuploidy is the duplication or loss of individual chromosomes or parts of chromosomes. Plants tolerate aneuploidy better than animals. When regions of DNA are duplicated that contain genes, several possible outcomes that may occur: the duplicated gene may lose function, gain a novel function, or the function may become partitioned. Genomes may also be rearranged by moving gene locations within a chromosome or by the fusion of two chromosomes, which produces similar effects to aneuploidy. Noncoding DNA may also be affected and acquire regulatory function. Horizontal gene transfer or gene swapping between species is also possible, which complicates phylogenies and gives rise to many questions.

Essential Knowledge covered
4.C.1: Variation in molecular units provides cells with a wide range of functions.

Review It

Determine if the following statements on genome evolution are true or false **(T/F)**:

Genome evolution only occurs through whole-genome duplication.

Aneuploidy is the duplication of the entire genome.

A duplicated gene may gain a novel function.

It is possible for two chromosomes to fuse together.

Use It

Icefish live in Antarctica in water that is almost two degrees colder than water in which most other fish freeze solid. A genome study found the icefish's unique abilities can be traced to changes in gene that normally codes for pancreatic digestive protein. Suggest why and how the gene may now allow for icefish to survive in freezing water.

24.4 Gene Function and Expression Patterns

Recall It

While many species may share genetic similarities at the root of their DNA, the way in which their genes are expressed can lead to vast differences in phenotype. Even when species have highly similar genes, expression of these genes may vary greatly.

Essential Knowledge covered
1.B.1: Organisms share many conserved core processes and features that evolved and are widely distributed among organisms today.

Review It

Recalling different molecular techniques you read about in Chapter 18, what tool would be best at determining differences in gene transcription between two different species?

Use It

Human and chimpanzee genomes are 98.77% identical overall, and have 99.2% of the same exact genes. Why are chimp and human phenotypes so different, and considered separate species?

24.5 Applying Comparative Genomics

Recall It

Comparative genomics can be applied to many different areas of biology. By comparing related organisms, researchers can focus on genes that cause diseases and devise possible treatments. Analysis of the genome targets of pathogenic organisms may provide new avenues of treatment and prevention. Comparative genomics can also be applied to population of endangered species so that conservations might help reduce disease and develop breeding programs.

Review It

A study found that three different species of protozoans responsible for three different diseases shared over 6000 genes in common. How might this information be useful to scientists trying to develop a single treatment to cure all three diseases?

24 **Chapter Review**

Summarize It

How do similarities and differences in the genomes of chimpanzees and humans reveal about their evolutionary ancestry? Use the many different pieces of data provided by comparative genomics and discussed in Chapter 24 to help answer the question.

Chapter 25: Evolution of Development

Extending Knowledge

In Part III and Part IV of this textbook, you have become familiar genes, gene expression, and with the process of evolution. An understanding of development draws on all of this previously learned information. While the information in Chapter 25 is not required for the AP exam, it does provide examples on how natural selection, cellular communication, and gene expression impact the evolution of development.

Big Idea 1

Big Idea 3

Chapter Overview

Genes have evolved over time to produce the vast phenotypic differences that we see in species today. As you learned in Chapter 9, many signals mediate gene expression. *Hox* genes are an example of highly conserved genes found in plants and animals. Differences in shoot growth and development in plants and body morphology in animals depend on when and how these genes are expressed. Complex regions of the nervous system, such as vision (which you will read more about in later chapters) can be traced to a branch of genes called *Pax* genes. It is thought that independent diversification of *Pax* genes in descendants from an ancestor with a simple light receptor gradually lead to different, sophisticated, and complex eyes.

Review It

Determine if the following statement is true or false **(T/F)**:

Changes in DNA can never lead to phenotypic differences.

Modifying the location of gene expression can lead to changes in phenotypic differences.

Single-gene mutations can lead to rapid evolutionary change.

Convergent structures have the same or similar function but arose independently.

Summarize It

If the *Hox* genes, *Hoxc* and *Hoxd*, in mice were both altered so they could no longer bind to their target genes that promote digit formation, what phenotypic change would you expect to see and why?

Compare and contrast the eyespots found in ribbon worms to the eyes of a mouse.

Chapter 26: The Origin and Diversity of Life

Essential Knowledge

1.A.4	Biological evolution is supported by scientific evidence from many disciplines, including mathematics. **(26.1)**	Big Idea 1
1.D.1	There are several hypotheses about the natural origin of life on Earth, each with supporting scientific evidence. **(26.2)**	Big Idea 1
1.D.2	Scientific evidence from many different disciplines supports models of the origin of life. **(26.1, 26.2, 26.3, 26.5)**	Big Idea 1
2.A.2	Organisms capture and store free energy for use in biological processes. **(26.4)**	Big Idea 2

Chapter Overview

Strong evidence supports the hypothesis that all life on Earth descended from a common ancestor. There are several theories regarding the origins of life on Earth. All theories, however, agree that over Earth's history life has greatly diversified. This chapter explores some of the hypotheses on the origins of life and describes how Earth's geological history led to life's diversification.

26.1 | Deep Time

Recall It

Earth has changed dramatically over its 4.6 billion year history. Earth has experienced extreme shifts in temperature that correspond with shifting CO_2 levels. Continents have formed, collided to make supercontinents, and separated again multiple times. Life on Earth has only occurred in the last 12% of Earth's history, and humans have only existed for 0.2% of Earth's history. The fossil record provides clues to help us reconstruct our biological past.

Essential Knowledge covered
1.A.4: Biological evolution is supported by scientific evidence from many disciplines, including mathematics.
1.D.2: Scientific evidence from many different disciplines support models of the origin of life.

Review It

Identify the following geological terms that are helpful in understanding the history and distribution of life on Earth:

Description	Geological Term
The movement of rigid slabs of rock under continents and oceans	
Evidence of early life captured in rocks	
Gas added to the atmosphere through volcanic eruptions	

Use It

The study of life is what defines biology, but biology draws regularly on other scientific fields. Explain how evidence from a field other than biology contributes to our understanding of the history of life on Earth.

26.2 Origins of Life

Recall It

Life most likely evolved in Earth's early chemically rich oceans. The key organic molecules for life may have been formed in the reducing atmosphere of early Earth. Biochemists have attempted to recreate Earth's early conditions and have found biologically important molecules can be formed under such conditions. There is also evidence that organic molecules may have extraterrestrial origins, as many amino acids have also been found on meteorites. Metabolic pathways and membranes evolved early, forming single-celled organisms.

Essential Knowledge covered
1.D.1: There are several hypothesis about the natural origin of life on Earth, each with supporting scientific evidence.
1.D.2: Scientific evidence from many different disciplines support models of the origin of life.

Review It

Name the four macromolecules that are the building blocks of life.

Determine if the following statement applies to RNA **(R)** or DNA **(D)**:

   Is probably the first nucleic acid that allowed for self-replication

   Is made of ribose

   Is more stable

Use It

Describe the Miller-Urey experiment. What was the hypothesis? How was it tested?

List two pieces of evidence that support the hypothesis that Earth was once an RNA world.

Recall It

The earliest fossil dates back to approximately 3.8 BYA and are microscopic single-celled organisms. Carbon dating revealed that such organisms sequestered carbon using an ancient form of photosynthesis. Organic molecules that clearly have biological origins, including lipids, have been identified in 2.7-billion-year-old rock formations in Australia.

Essential Knowledge covered
1.D.2: Scientific evidence from many different disciplines support models of the origin of life.

Review It

Recall at least three of the seven characteristics shared by all living things (see Chapter 1 for review).

Use It

Describe at least two pieces of evidences that supports the ideas that life on Earth dates back to as early as 3.8 BYA.

26.4 Earth's Changing System

Recall It

Since its formation, Earth's climate has been changing. Extreme shifts in temperature have correlated with changes in CO_2 levels and with glaciation events. At least three glaciation events have covered the entire globe, leading to mass extinction and affecting the course of evolution. Shifts in the atmosphere are the result of geologic changes and changes in the types and abundance of living organisms. Geological weathering in tropical climates and on the exposed surfaces of newly formed continents decreased atmospheric CO_2. In some cases, the drop in CO_2 concentrations can be sufficient to trigger glaciations. Continental motion also affects evolution through creating or destroying geographical barriers. The evolution of photosynthetic organisms dramatically changed Earth's climate, as well as terrestrial evolution, through the production of O_2 and indirect production of the ozone layer. Plant life may also be responsible for two glaciation events through drawing down CO_2 levels from the atmosphere.

Essential Knowledge covered
2.A.2: Organisms capture and store free energy for use in biological processes.

Review It

Explain the phenomenon of Snowball Earth. How does this help to explain the increase or decrease in diversity?

How many glaciation events occurred in Earth's history?

Use It

Describe one way in which the evolution of plants aided in the evolution of land animals, and one way the presence of plants on Earth hindered land animals.

26.5 Ever-Changing Life on Earth

Recall It

Several key evolutionary events support the diversification of life in response to an ever-changing environment. Compartmentalization, multicellularity, and sexual reproduction established the foundations for the rapid evolution of animal life during the Cambrian period. Compartmentalization of cells and the acquisition of mitochondria and chloroplasts allowed for multicellularity. Multicellularity arose independently in eukaryotic supergroups. The evolution of meiosis and sexual reproduction increased the genetic diversity available for natural selection.

Essential Knowledge covered
1.D.2: Scientific evidence from many different disciplines support models of the origin of life.

Review It

List two structures subdivide a cell into different functional compartments.

Use It

What was the Cambrian radiation? How do we know so much about it?

Summarize It

The following graph shows an estimate of geological oxygen levels in the atmosphere. Where did the increasing levels of O_2 come from, and why was it important for the development of life on Earth?

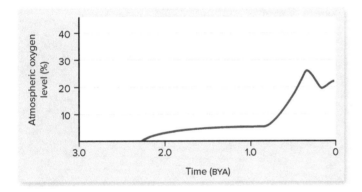

Chapter 27: Viruses

Essential Knowledge

3.A.1	DNA, and in some cases RNA, is the primary source of heritable information. (27.1)	Big Idea 3
3.C.3	Viral replication results in genetic variation, and viral infection can introduce genetic variation into the hosts. (27.1, 27.2, 27.3, 27.4)	Big Idea 3

Chapter Review

Viruses are genetic elements enclosed in protein with disease-producing potential. Diseases caused by viruses can be lethal to many types of organisms. Viruses are also used experimentally in biotechnology to carry genes from one organism to another.

27.1 The Nature of Viruses

Recall It

The basic structure of a virus consists of either DNA or RNA encased in a protein coat called a capsid. Animal viruses may also have an envelope around the capsid composed of proteins, lipids, and glycoproteins. The viral genome may be linear or circular, and single- or double-stranded. Viruses are classified as DNA viruses, RNA viruses, or retroviruses. RNA viruses may contain multiple RNA molecules or only one. Retroviruses contain RNA that is transcribed into DNA by reverse transcriptase. Viruses are obligate intracellular parasites, lacking the ribosomes and proteins needed for replication. Instead, viruses replicate and take over host machinery while directing their own nucleic acid and protein synthesis. Viruses vary in size and come in two simple shapes: helical or icosahedral.

Essential Knowledge covered
3.A.1: DNA, and in some cases RNA, is the primary source of heritable information.
3.C.3: Viral replication results in genetic variation, and viral infection can introduce genetic variation to the hosts.

Review It

Draw a picture of an icosahedral DNA bacterial virus and a helical RNA animal virus. Be sure to label the nucleic acid and capsid and any other structures of importance.

Describe the following viral structures:

Structure	Description
Capsid	
Nucleic acid	
Envelope	

Use It

Can a virus replicate outside a host? Why or why not?

Why is it difficult to make a vaccine for a RNA virus?

27.2 Bacteriophage: Bacterial Viruses

Recall It

Bacterial viruses are called bacteriophage. Bacteriophage may have two reproductive cycles: the lytic cycle and the lysogenic cycle. In the lytic cycle, the phage attaches to a host cell, penetrates the cell membrane, and synthesizes more viruses. The virus particles are released as the cell membrane ruptures or buds, killing the host cell. The lysogenic cycle incorporates the virus's genetic material into the host genome as a prophage. The prophage may lie dormant and be replicated until induced by DNA damage or other environmental damage, causing the viral particles to enter the lytic cycle. Bacteriophage can also contribute genes to the host genome.

Essential Knowledge covered
3.C.3: Viral replication results in genetic variation, and viral infection can introduce genetic variation to the hosts.

Review It

Determine if the following statements apply to either the lytic or the lysogenic cycle:

Can establish a latent infection

Is replicated alongside the host DNA as the cell divides.

Can be stimulated via external stressors such as UV radiation

Use It

Create a diagram of the lytic cycle of a virus in a cell showing the five major steps of viral replication. Place a star by the step in which the virus may exit to the lysogenic cycle or where it may exit the lysogenic cycle and return to the lytic cycle.

Recall It

The human immunodeficiency virus (HIV) causes the disease called acquired immunodeficiency syndrome (AIDS). HIV is a well-studied virus, which compromises the host immune system by targeting macrophages and helper T-lymphocyte cells. With the loss of these cells, the body cannot fight off opportunistic infections, which ultimately leads to death. The glycoproteins on HIV target the receptors on macrophages and T cells, activating endocytosis. Once inside, viral RNA is released, and it undergoes reverse transcriptase. HIV has a high mutation rate because the reverse transcriptase enzyme is much less accurate than DNA polymerases. Mutations lead to altered glycoproteins, which can bind to different receptors. AIDS treatments target different phases of the HIV life cycle including reverse transcriptase, a protease involved in protein maturation, viral entry, and the integration of the genome. Vaccine development for HIV has been unsuccessful.

Essential Knowledge covered
3.C.3: Viral replication results in genetic variation, and viral infection can introduce genetic variation to the hosts.

Review It

Describe the following stages of HIV infection:

Stage	Description
Attachment	
Entry into cell	
Replication	
Assembly	

Use It

Describe two reasons that HIV is a persistent and dangerous virus.

Recall It

The most lethal flu virus in human history is the Type A influenza virus, which can also infect other mammals and birds. There are also Type B and Type C flu viruses that are restricted to humans and rarely cause health problems. Influenza viruses undergo genetic recombination frequently, so they are not recognized by antibodies against past infections. Each year the composition of flu vaccines must be changed. New viruses emerge by infecting new hosts. Viruses can extend their host range by jumping to another species. Examples of emergent diseases include hantavirus, hemorrhagic fever, and SARS. Viruses have also been linked to formation of cancers, including liver cancer and cervical papillomas.

Essential Knowledge covered
3.C.3: Viral replication results in genetic variation, and viral infection can introduce genetic variation to the hosts.

Review It

Determine whether the following statements are true or false **(T/F)** about viruses:

Development of flu vaccines are easy; recombination is not common in the influenza virus.

Viruses may contribute to 15% of human cancers.

Emerging viruses originate in one organism and then are passed to another species.

Type B influenza virus is the most lethal in human history and has caused several pandemics.

New strains of flu frequently originate in Asia.

Use It

Describe why Ebola and SARS are considered emerging viruses.

Recall It

Prions and viroids are infectious particles that are smaller and simpler than viruses. Prions contain a misfolded form of a protein and no nucleic acids. Prions catalyze a chain reaction of protein misfolding. Viroids are circular molecules of RNA that can infect plants and use host proteins to replicate.

Review It

Why is a viroid not technically a virus?

How do prions cause neurodegenerative diseases?

---

27 Chapter Review

Summarize It

Using the diagram below, explain how the swine flu (H1N1) is related to avian flu (H5N1).

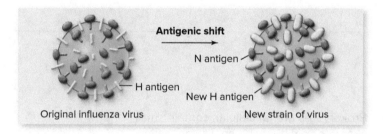

Chapter 28: Prokaryotes

Essential Knowledge

1.A.2	Natural selection acts on phenotypic variations in populations. **(28.3)**	**Big Idea 1**
2.A.2	Organisms capture and store free energy to use in biological processes. **(28.4)**	**Big Idea 2**
2.B.1	Cell membranes are selectively permeable due to their structure. **(28.1)**	**Big Idea 2**
2.D.1	All biological systems from cells and organisms to populations, communities and ecosystems are affected by complex biotic and abiotic interactions involving exchange of matter and free energy. **(28.5, 28.6)**	**Big Idea 2**
3.C.2	Biological systems have multiple processes that increase genetic variation. **(28.3)**	**Big Idea 3**
3.C.3	Viral replication results in genetic variation, and viral infection can introduce genetic variation into the hosts. **(28.3)**	**Big Idea 3**
4.A.6	Interactions among living systems and with their environment result in the movement of matter and energy. **(28.6)**	**Big Idea 4**

Chapter Overview

Prokaryotes are unicellular organisms and are the dominant life forms on Earth. Prokaryotes include the domains bacteria and archaea.

28.1 Prokaryotic Diversity

Recall It

Prokaryotes differ from eukaryotes in several key ways. Prokaryotes are unicellular, contain small circular DNA, and lack internal compartmentalization. Prokaryotes divide by binary fission, have only a singular flagellum, and are metabolically diverse. While both bacteria and archaea have prokaryotic structure, they differ in four key areas: plasma membranes, cell walls, DNA replication, and gene expression. The composition of plasma membranes differs in bacteria and archaea. The cell walls of bacteria contain peptidoglycans, but those of archaea do not. Both bacterial and archaeal DNA have a single replication origin, but the origin and the replication proteins are different. Archaeal DNA replication is more like that of eukaryotes. Most bacteria are capable of forming complex communities of many different species known as biofilms. There are nine clades of prokaryotes have been found so far, but many bacteria have not been studied.

Essential Knowledge covered
2.B.1: Cell membranes are selectively permeable due to their structure.

Review It

Identify the domain of life based on the characteristics provided:

Domain	Characteristics
	Multicellular. Contains compartmentalized cells with linear DNA.
	Unicellular. Cell walls lack peptidoglycan. DNA replication is similar to eukaryotes.
	Unicellular. Cell walls contain peptidoglycan. Plasma membranes contain ester linkages.

Use It

Describe the approach you would take to identify different bacteria in a biofilm.

28.2 Prokaryotic Cell Structure

Recall It

Prokaryotes are often categorized based on their cell shape. There are three basic shapes: rods, cocci, and spirals. Prokaryotic cell walls are tough. Some bacteria may have a gelatinous layer called a capsule that enables the bacterium to adhere to surfaces and evade an immune response. Many bacteria have a slender flagellum composed of flagellin, which can rotate to drive movement. Other bacteria have hair-like pili that may play a role in adhesion or in the exchange of genetic information. Many prokaryotes form endospores, which are highly resistant to environmental stress and allow the prokaryote to germinate when conditions become favorable. Prokaryotes have a simple interior structure, consisting of internal membranes, a nucleoid region, and ribosomes.

Review It

Compare and contrast flagella and pili.

Recall It

Conjugation is the transfer of genes from one cell to another cell through cell-to-cell contact. Bacteria that contain F plasmids are called F$^+$ cells. F$^+$ cells may pass an F plasmid to an F$^-$ cell via a conjugation bridge. Viruses transfer DNA by transduction. There are two types of transduction: generalized and specialized. Generalized transduction occurs when viruses package host DNA and transfer it on subsequent infection. Specialized transduction is limited to lysogenic phages and only a few genes are transferred. Transformation is the uptake of DNA by bacteria directly from the environment, and can be artificially induced. Antibiotic resistance can be transferred by resistance plasmids, known as R plasmids. R plasmids have played a significant role in the appearance of strains resistant to antibiotics, such as *S. aureus* and *E. coli* O157:H7. Genetic variation in prokaryotes may also arise through spontaneous mutations due to exposure to radiation, UV, and various chemicals.

Essential Knowledge covered
1.A.2: Natural selection acts on phenotypic variations in populations.
3.C.2: Biological systems have multiple process that increase genetic variation.
3.C.3: Viral replication results in genetic variation, and viral infection can introduce genetic variation to the hosts.

Review It

Identify the process of genetic transfer in prokaryotes:

Description	Process
Uptake of naked DNA from the environment	
Viral transmission of DNA to bacteria	
Cell-to-cell transfer of DNA	

List three different environmental factors that may lead to mutations in bacterial DNA.

Why is *Staphylococcus aureus* with an R plasmid of concern to humans?

˙ a rise in antibiotic resistant bacterial infections both in and out of
ᴺntibiotic resistant bacteria come from?

Compare and contrast generalized transduction with specialized transduction.

28.4 Prokaryotic Metabolism

Recall It

Prokaryotes have diverse mechanisms to acquire energy and carbon for growth and reproduction. Autotrophs obtain their carbon from inorganic CO_2 whereas heterotrophs need at least some carbon from organic molecules. Prokaryotic metabolism can be categorized into the following four groups: (1) photoautotrophs, (2) photoheterotrophs, (3) chemolithautotrophs, and (4) chemoheterotrophs. Photoautotrophs carry out photosynthesis and obtain carbon from carbon dioxide. Photoheterotrophs also use light for energy but obtain carbon from organic molecules. Chemolithoautotrophs obtain energy by oxidizing inorganic substances such as ammonia or sulfur. Chemoheterotrophs, the largest group, obtain carbon and energy from organic molecules. This group is comprised of decomposers and pathogens.

Essential Knowledge covered
2.A.2: Organisms capture and store free energy for use in biological processes.

Review It

Review the examples provided to you in this chapter to identify the metabolic process the following prokaryotes perform:

Prokaryote Name	Metabolic Category
Decomposers	
Nitrifies	
Cyanobacteria	
Pathogens	
Purple and green nonsulfur bacteria	

Use It

The ocean floor is completely devoid of sunlight. Instead, the water is rich with sulfides from thermal vents and the carcasses of dead plants and animals. What categories of prokaryotes would you expect to find here? What categories would be absent? Justify your answers.

28.5 Human Bacterial Disease

Recall It

Bacteria are responsible for many human diseases. Bacterial diseases can be spread through numerous vectors, including mucus or saliva droplets, contaminated food and water, and insects. Examples of bacterial borne diseases include tuberculosis, which is caused by *Mycobacterium tuberculosis*, and affects the respiratory system. Some bacterial can cause tooth decay, stomach ulcers, and a number of sexually transmitted diseases, including gonorrhea, syphilis, and chlamydia.

Essential Knowledge covered
2.D.1: All biological systems from cells and organisms to populations, communities and ecosystems are affected by complex biotic and abiotic interactions involving exchange of matter and free energy.

Review It

Determine if the following statements concerning human bacterial disease are true or false **(T/F)**:

Tuberculosis is easily transmitted from person to person through the air.

Chlamydia causes symptoms in its host immediately.

Diseases caused by bacterial biofilms are difficult to treat.

The majority of peptic ulcers are caused by a bacterial infection.

Most sexually transmitted diseases are viral in origin and not bacterial.

Use It

The graph to the right shows the number of cases of three different sexually transmitted diseases in the United States over a 20 year period. Provide an explanation for why chlamydia cases may increasing.

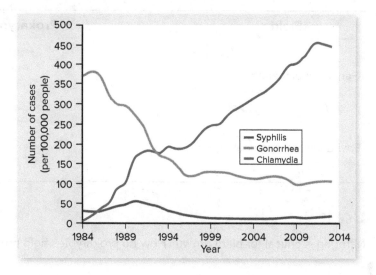

28.6 Beneficial Prokaryotes

Recall It

Prokaryotes carry out a number of critical processes. Prokaryotes are involved in the cycling of chemical elements between organisms and in the physical environment. Prokaryotes decompose dead organisms, returning elements back to Earth. Other prokaryotes fix or make biologically available nutrients such as a carbon and nitrogen. Prokaryotes often live in symbiotic associations with eukaryotes. Humans have genetically engineered prokaryotes to produce pharmaceuticals and other useful products. Bacteria are also used in bioremediation to remove harmful pollutants from the water, air, and soil.

Essential Knowledge covered
2.D.1: All biological systems from cells and organisms to populations, communities and ecosystems are affected by complex biotic and abiotic interactions involving exchange of matter and free energy.
4.A.6: Interactions among living systems and with their environment result in the movement of matter and energy.

Review It

List two biogeochemical cycles in which prokaryotes play an important role.

Provide a description of how humans use prokaryotes in the following engineering fields:

Field	Prokaryote Use
Genetic engineering	
Environmental engineering	

Use It

Nitrogen is critical for plant growth. How do prokaryotes help provide nitrogen to plants?

28 Chapter Review

Summarize It

1. A virus attaches to a bacterium and inserts a piece of bacterial DNA into the bacterium's cytoplasm. The DNA belongs to another bacteria species but is incorporated into the new bacterium's genome. How could the virus contain bacterial DNA?

2. A hospital suddenly began experiencing an outbreak of methicillin-resistant *Staphylococcus aureus* (MRSA) infections. Design a plan to collect data to determine how the outbreak may have started.

3. Which type of prokaryote would be negatively affected by decreasing levels of carbon dioxide in the atmosphere: a photoautotrophs or a photoheterotroph, and why?

Chapter 29: Protists

AP

Extending Knowledge

In Part IV, you learned that organisms share many conserved processes which support the relatedness within phylogenetic groups. Chapter 29 introduces protists, one of the kingdoms of eukaryotes, and discusses examples of structural and functional features shared by eukaryotes.

Big Idea 1
Big Idea 2

Chapter Overview

Protists were the first eukaryotic organisms to evolve. As you first learned in Chapter 4, eukaryotic cells are thought to have arisen through endosymbiosis. Protist have diverse metabolisms, structures, and life cycles. Protist play significant roles in ecosystems, including producing oxygen and the foundations of the food web in aquatic systems.

Review It

Draw a diagram of a eukaryotic cell. Include in your drawing: a nucleus, mitochondria, the endoplasmic reticulum, a golgi apparatus, and some chloroplasts. Place a star by the two organelles that are thought to have evolved from endosymbiosis (it may be helpful for you to review Chapter 4).

Summarize It

Red and brown algae are protists that have chloroplasts. Red algae only have two membranes surrounding the chloroplasts, while brown algae have four membranes. Using the theory of endosymbiosis, determine how these protists evolved to have different numbers of membranes around their chloroplasts. Use a diagram to help explain your answer.

The protist *Euglena* has a flask shaped opening called a reservoir at the anterior end of the cell. Contractile vaucoles collect excess water from all parts of the organism and empty into the reservoir. What use do you suppose this mechanism is to the organism?

Chapter 30: Seedless Plants

Extending Knowledge

In Part IV, you learned that organisms share many processes which reflect both common ancestry as well as differences due to adaptations in different habitats. Chapter 30 introduces land plants, which exhibit both shared properties with freshwater algae as well as unique osmoregulatory mechanisms for terrestrial environments.

Big Idea 2

Chapter Overview

All land plants evolved from green algae and are adapted to life outside of the water. The first land plants lacked seeds and included bryophytes, tracheophytes, lycophytes, and pterophytes. As you learned in Chapter 2, water vital to life, and its unique properties drive many biological processes. Once plants moved onto land, they evolved many different mechanisms to maintain the necessary amounts of water in their tissues. Osmoregulation, the processes in which an organism maintains water balance, is particularly important to land plants.

Review It

List four properties of water that make it unique, and how these properties are vital for life.

Explain the property of cohesion in water molecules. Why might this property be particularly important to land plants?

Summarize It

Bryophytes, which contain seedless plants such as mosses, liverworts, and hornworts, were some of the first land plants to evolve. Why are these plants are relatively small in size?

Land plants have a waxy cuticle on above-ground parts and pores called stomata. How do these features play a role in plant osmoregulation?

Chapter 31: Seed Plants

Essential Knowledge

2.E.1	Timing and coordination of specific events are necessary for the normal development of an organism, and these events are regulated by a variety of mechanisms. **(31.1, 31.3, 31.4)**	**Big Idea 2**

Chapter Review

Seed-producing plants are highly successful land plants that are incredibly diverse. The evolution of the seed allowed land plants to become highly dispersed, as embryos were protected and could germinate only under favorable conditions. Seed-producing plants have become fundamental to every terrestrial biome and to human existence.

31.1 The Evolution of Seed Plants

Recall It

Seed-producing plants, the gymnosperms and angiosperms, evolved from spore-bearing plants. Seeds offer more protection than spores by having an extra layer of sporophyte tissue, called the integument, surrounding the embryo. The seed keeps the embryo from drying out and provides stored food, the endosperm, for the embryo. The seed also allows for a dormant stage that pauses the life cycle until environmental conditions are favorable.

Essential Knowledge covered
2.E.1: Timing and coordination of specific events are necessary for the normal development of an organism, and these events are regulated by a variety of mechanisms.

Review It

Label the parts of the seed.

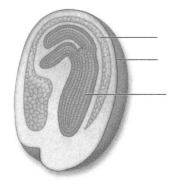

Use It

Describe two ways in which a seed protects the embryo of a developing plant.

31.2 Gymnosperms: Plants with "Naked Seeds"

Recall It

Gymnosperms are seed-plants that mostly produce cones. Gymnosperms encompass the groups conifers, cycads, gnetophytes, and ginkgophytes. Unlike angiosperms, gymnosperms do not produce flowers or fruits. The gymnosperms are characterized by their ovules not being fully enclosed by sporophyte tissues when they are pollinated.

Review It

Determine whether the following statements are true or false **(T/F)** about gymnosperms:

Gymnosperms produce fruit in the summer.

Gymnosperms produce two kinds of gametophytes: male and female.

Gymnosperms produce seeds.

Gymnosperms include palm trees.

Gymnosperms produce flowers.

31.3 Angiosperms: The Flowering Plants

Recall It

Unlike gymnosperm, angiosperms have ovules completely enclosed within diploid tissue called the ovary at the time of fertilization. Angiosperms also form flowers and fruits. Flowers are considered modified stems that bear modified leaves and house the gametophyte. Many flowers attract pollinators through the production of nectar. Pollinators, such as insects, birds, and other animals, mechanically transfer pollen from flower anthers to flower stigmas. Pollination can also be triggered by wind, water, and gravity.

Essential Knowledge covered
2.E.1: Timing and coordination of specific events are necessary for the normal development of an organism, and these events are regulated by a variety of mechanisms.

Review It

Determine if the following statement applies to angiosperms, gymnosperms or both:

Statement	Seed Plant
Contains conifers, cycads, ginkgoes, and gnetophytes	
Has seeds	
Ovules are completely enclosed and become fruit	
May rely on pollinators to aid in reproduction	
Ovules are not completely enclosed by sporophyte tissue	

Use It

The evening primrose is an angiosperm aptly named because its flowers open at dusk and remain opening throughout the night. What type of pollinators would you expect visit the evening primrose? Can you come up with an evolutionary hypothesis of how the primrose came to open at night?

31.4 Seeds

Recall It

As you first encountered in Section 31.1, seeds protect the embryo in a number of ways. Seeds protect the embryo, provide a means for dispersal, and provide food for the embryo. Another important way seeds help to ensure the survival of the next generation is by maintaining dormancy during unfavorable conditions. As a seed coat must become permeable so that water and oxygen can reach the embryo before it can germinate. There are many adaptations seen in seed plants that evolved to ensure germination under appropriate survival conditions. In certain gymnosperms, seeds may be released from cones after a fire. Alternatively, seeds may require passage through a digestive tract, freeze–thaw cycles, or abundant moisture.

Essential Knowledge covered
2.E.1: Timing and coordination of specific events are necessary for the normal development of an organism, and these events are regulated by a variety of mechanisms.

Review It

List four ways seeds are an important to seed plant survival.

Use It

Seeds are a very successful adaptation in plants, allowing angiosperms and gymnosperms to become the dominant plants on the planet. Describe an adaptation seen in seed plants that evolved to ensure germination under appropriate survival conditions.

Recall It

Fruits are the mature ovaries of angiosperms. Angiosperms produce many different types of fruits; dry or fleshy, and simple, aggregate, or multiple. Fruits are helpful for dispersal, allowing angiosperms to colonize large areas. Many fruits are ingested and transported away by animals. Some fruits are buried by herbivores, blown away by the wind, and some even float on water.

Review It

Compare the fruit dispersal mechanism of an oak tree that produced acorns with a dandelion. Which do you think would have a wider distribution and why?

31 **Chapter Review**

Summarize It

Jack pines are gymnosperms that require the high temperature from fire to release seeds from their cones. If an area with a relatively small population of jack pines experienced a large forest fire, and most of the adults were destroyed, would the resulting populations of jack pines in the following years be smaller or larger? Explain your answer.

Chapter 32: Fungi

Extending Knowledge

Big Idea 2

In Chapter 22, you learned that internal and external signals help organisms respond to environmental conditions and cues. The fungi discussed in Chapter 32 exhibit multiple examples of this principle, using multiple mechanisms to regulate timing of important events such as reproduction.

Chapter Overview

Fungi are heterotrophic organisms that obtain their nutrients through excreting enzymes for external digestion and then absorbing the products. Fungi contain chitin in their cell walls, and are more closely related to animals than to plants. As you were first introduced to in Chapter 31, reproductive events are regulated by multiple mechanisms and may require certain environmental cues in order to occur. Reproduction in fungi is no exception. Fungi usually produce spores for either sexual or asexual reproduction. Spores may be dispersed by the wind or may only be expelled when exposed to other factors. Some fungi can reproduce asexually by budding. Fungi have a large impact in human lives from the yeast that make bread and beer to the fungi that can cause disease in animals and plants. Fungi also play an important role in the ecosystem through forming important symbioses with plants and animals.

Review It

List three unique characteristics of fungi.

Define sexual reproduction, and compare it to asexual reproduction.

Describe the different forms of symbiosis listed below:

Symbiosis	Description
Parasitism	
Facultative symbiosis	
Mutualism	
Obligate symbiosis	
Commensalism	

Summarize It

Some fungi reproduce sexually when environmental conditions are unfavorable. Why would this mode of sexual reproduction be beneficial when environmental conditions are changing?

One summer morning, following an evening of warm rain, you notice that your front yard is full of neat white mushrooms with red caps, which seemingly popped up out of nowhere! Using what you have learned in this chapter, provide a biologically sound explanation of where these mushrooms may have come from and why they suddenly appeared.

The phylum of fungi Ascomycota contains the fungi *Ophostoma ulmi*. *O. ulmi* is carried to American elm trees through elm bark beetles. Elm bark beetles lay their eggs in crevices in the trees. The beetle larvae that hatch then form small tunnels throughout the tree as they grow. Given what you have learned about how fungi grown and obtain nutrients, explain why *O. ulmi* is ultimately responsible for the deterioration and death of American elm trees.

Review section 32.8 in your textbook. Describe two mutualistic relationships involving fungi.

Chapter 33: Animal Diversity and the Evolution of Body Plans

Extending Knowledge

In Part IV, you learned that organisms share many processes which reflect both common ancestry as well as differences due to adaptations in different habitats. Chapter 33 provides an overview of how the body plans of animals diversified and evolved throughout time.

Chapter Overview

Animals are a diverse group in shape and size, but they share many core characteristics such as heterotrophy, multicellularity, lack of cell walls, and active movement. Five key evolutionary innovations are found in animals: (1) symmetry, (2) tissues that allow for specialized structure and function, (3) a body cavity, (4) various patterns of embryonic development, and (5) segmentation. There are many phyla of animals, but this chapter focused on the earliest animals to evolve: Parazoa (animals such as jellyfish, worms, and sponges) and Eumetazoa (animals such as hydras, anemones, and corals). The Bilateria (protostomes and deuterostomes) are introduced here and discussed in more depth in Chapter 34 and 35.

Review It

Identify the system that distributes nutrients and oxygen throughout an animal's body.

Identify the system that removes carbon dioxide and water from an animal's body.

Animals are heterotrophs. Where do they get their organic compounds?

Summarize It

Sponges and cnidarians have very basic body plans as described in sections 33.4 and 33.5. Compare and contrasts how these two organisms obtain nutrients, digest food, and deal with waste products.

Why are sponges and cnidarians considered "primitive" animals?

Chapter 34: Protostomes

Extending Knowledge

In Part IV, you learned that organisms share many processes that reflect both common ancestry as well as differences due to adaptations in different habitats. Chapter 34 provides an overview of the evolution of protostomes, and how different clades adapted to a diversity of environments.

Chapter Overview

Protostomes are bilateralian animals that develop in a specific way: the mouth of the adult animal develops from or near the blastopore. Protostomes exhibit great diversity in digestive, circulatory, and reproductive systems, not to mention morphology. Protostomes include a number of different organisms from flatworms, rotifers, molluscs, ribbon worms, annelids, bryozoans, round worms, and arthropods.

Review It

Sort the following protostomes into the clade Spiralia or Ecodysozoa: rotifer, round worm, mollusk, flatworms, bryozoan, arthropod

Spiralia	Ecodysozoa

Summarize It

Use the information provided to you in section 34.2 to describe how a flatworm digests the food it ingests. How is this different from a tapeworm?

From the outside, nemerteans, which include ribbon worms, appear similar to flatworms. Internally, however, they are quite different. Using section 34.5, explain some of these differences.

Squid are members of the phylum Mollusca. Referring to section 34.4, draw a picture of the basic squid body plan. Label the structures that aid in the exchange of gases, circulation of fluids, digestion of food, and excretion of waste.

Chapter 35: Deuterostomes

Extending Knowledge

In Part IV, you learned that organisms have specialized areas or body systems that perform specific functions, and that this specialization allows organisms to more efficiently process energy and matter. Chapter 35 provides an overview of the evolution and specialization of deuterostomes, and how different clades developed specializations that increased their survival.

Chapter Overview

Deuterostomes are eukaryotes that exhibit the same pattern of development, in which the anus develops from the blastopore and the mouth develops from another part of the embryo. There are two major phyla of deuterostomes: Echinodermata and Chordata. Echinoderms have a water-vascular system and tube feet that aid in movement and feeding. You may be most familiar with the echinoderm called the sea star. Members of the phylum Chordata have a nerve cord, notochord, pharyngeal slits, and a postanal tail at some point in their lives. Chordata is further subdivided into nonvertebrate and vertebrate organisms, which display a wide range of diversity in physiology and morphology.

Review It

Identify the chordates based on their description:

Chordate	Description
	Nonvertebrate marine chordates that lack bones and a distinct head as an adult.
	Vertebrates with jaws, paired appendages, internal gills, and a closed circulatory system.
	Nonvertebrates with a swimming larval that form and a sessile, baglike adult.
	Vertebrate with a watertight amniotic egg, watertight skin, and thoracic breathing.
	Vertebrate with hair, mammary glands, and are endothermic.
	Vertebrate endotherms with feathers, and a highly efficient respiration.
	Vertebrate with legs, lungs, cutaneous respiration, and a partially divided heart.

The organisms in the chart above are all share the following characteristics seen at some point during development: nerve cord, notochord, pharyngeal slits, and a postanal tail. What does this indicate about all chordates?

Summarize It

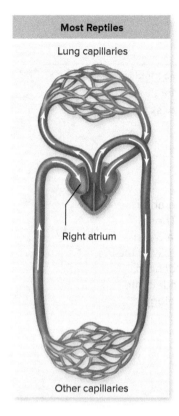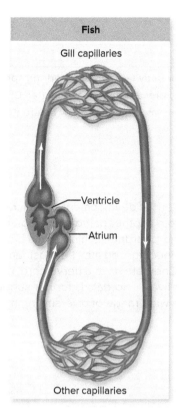

Most Reptiles

Lung capillaries

Right atrium

Other capillaries

Fish

Gill capillaries

Ventricle

Atrium

Other capillaries

Using the diagram provided, explain how the circulatory system of most reptiles differ from the circulatory system of fishes.

Compare and contrast how a bird and a reptile regulate their body temperatures.

Chapter 36: Plant Form

Extending Knowledge

In Part IV, you learned that organisms have specialized areas or body systems that perform specific functions, and that this specialization allows organisms to more efficiently process energy and matter. Chapter 36 provides an overview of the form and function of plants, and how specialization of features such as roots and leaves increased their survival.

Big Idea 2

Big Idea 4

Chapter Overview

The vascular plants you read about in Chapters 30 and 31 are incredibly diverse, but they all have the same basic structure. Plants have particular ways they grow, create, and transport food throughout their structures. They also have similarities in how they develop. Vascular plants are composed of roots and shoots. Roots are necessary for anchoring the plant in the soil, as well as for absorbing water and nutrients. Shoots, or stems, support the above ground organs, such as flowers and leaves. Plants have different tissues and specialized cells which are important for storage of carbohydrates, protection and support, as well as the movement of water and nutrients throughout the plant. Of particular importance to plants are the processes of transpiration and, of course, photosynthesis, which takes place in the leaves.

Review It

Recall the process of photosynthesis that you learned about in Chapter 8. What are the reactants? What are the products? Where does this process occur in vascular plants?

Plants have many specialized structures and tissues. Using this chapter, describe the purpose of the following structures and tissues:

Structure/Tissue	Purpose
Root	
Trichomes	
Phloem	
Stomata	
Root hairs	
Stem	
Xylem	
Leaf	

Summarize It

Using what you learned about plant structures in this chapter, as well as what you learned in Chapter 2 regarding the unique properties of water, draw a diagram and explain how water is transported from the soil to the leaves of a plant.

Carbon dioxide is critical for photosynthesis to occur. Explain how carbon dioxide is able to enter a plant.

What protection, if any, do plants have against predators?

Chapter 37: Transport in Plants

Essential Knowledge

| 2.B.1 | Cell membranes are selectively permeable due to their structure. **(37.1)** | Big Idea 2 |
| 2.B.2 | Growth and dynamic homeostasis are maintained by the constant movement of molecules across membranes. **(37.2, 37.6)** | Big Idea 2 |

Chapter Review

Water, minerals, and organic molecules are necessary components for plant growth and development. Plants move water into and through their tissues using osmosis and transpiration. Specialized cells and tissues aid in water and nutrient transport.

37.1 Transport Mechanisms

Recall It

Water potential (Ψ) is the sum of pressure potential and solute potential and predicts which way water will move in a plant. Water moves from an area of high water potential to an area of low water potential. Once water is in the plant, transpiration is the major pulling force for water transport. The unique properties of water, including cohesion, adhesion, and osmosis contribute to water movement. Aquaporins, water channels in plasma membrane, can help maintain water balance in a cell through osmosis.

Essential Knowledge covered
2.B.1: Cell membranes are selectively permeable due to their structure.

Review It

Write the equation for total water potential (Ψ_w) for a plant cell.

Identify the property or process responsible for the following descriptions of water movement:

Description	Property
Water molecules cling together and form a chain.	
Water moves into a root cell.	
Water molecules stick to xylem.	
Water moves from a cell with high water potential to a cell with lower water potential.	
Water and minerals are pulled up the xylem.	

Which direction do carbohydrates move in a plant?

Use It

If a plant cell has a pressure potential (Ψ_p) equal to 1 MPa and is a solution potential equal to 0.5 MPa, what is the total water potential of the cell?

If that plant cell was placed in a solution that had a total potential of 0.9 MPa, which way would the water moves? What structural components of the plant cell may increase the rate of osmosis?

37.2 Water and Mineral Absorption

Recall It

As you learned in Chapter 36, root hairs and mycorrhizal fungi are two ways plants can increase their surface area for absorption of water and minerals. There are three ways for water and minerals to enter the vascular tissue: (1) movement through the cell walls, (2) between plasmodesmata, and (3) through membrane transports. Before water and minerals make it into the endodermis, they reach Casparian strips. Casparian strips in the endoderm force water and minerals to move across the cell membranes, allowing selective flow of water and nutrients to the xylem.

Essential Knowledge covered
2.B.2: Growth and dynamic homeostasis are maintained by the constant movement of molecules across membranes.

Review It

Provide the definition for the following structures found in or on plants:

Structure	Definition
Root hairs	
Xylem	
Mycorrhizal fungi	
Plasmodesmata	

Use It

Place in order the following in terms of increasing water potential: shoots, soil, roots

Mineral concentrations are usually higher inside plants than in the water coming into the plant. Describe how these ions are taken up by root cells.

37.3 Xylem Transport

Recall It

When transpiration from leaves is low, root pressure increases, forcing water up to the leaves. This can result in guttation, where water is pushed out through special cells on leaves. If you have ever been up early in the morning and noticed dew on leaves, guttation may be partially responsible. Transpiration, however, remains the main force for driving water and minerals from the roots to the leaves out through stomata. Cavitation is a process that can stop water transport in plants. Cavitation occurs when xylem is broken or cut, introducing air bubbles, which disrupt the cohesion-adhesion flow of water.

Review It

Which of the following vascular cells are experiencing cavitation?

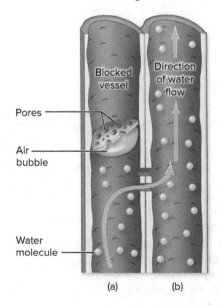

Describe the weather on a day you might observe guttation.

37.4 The Rate of Transpiration

Recall It

The rate of transpiration in a plant depends on many factors including weather, humidity, and time of day. Transpiration rates increase as temperature and wind velocity increase and as humidity decreases. Stomata are critical for maintaining the balance of water and carbon dioxide in a plant. Stomata open and close in response to changes in turgor pressure, changes in temperatures, and changes in carbon dioxide levels. Stomata open when the turgor pressure of guard cells increases due to the uptake of ions. Stomata close when guard cells lose turgor pressure and become flaccid. At high temperatures or when carbon dioxide concentrations increase, stomata close.

Review It

List three factors which may affect the rate of transpiration.

Determine if the following will make a stomata open **(O)** or close **(C)**:

External carbon dioxide levels increase

The sun shines on the stomata

External temperatures exceeds 34°C

Guard cells take up water

Abscisic acid is produced by wilting leaves

37.5 Water-Stress Responses

Recall It

Plants adaptations have evolved to allow plants to cope with environmental fluctuations such as drought, flood, and exposure to salinity. Plants that live in wet environments exhibit a variety of strategies to deal with constant inundation of water, including lenticels, adventitious roots such as pneumatophores, and aerenchyma tissue to ensure oxygen for submerged parts. For plants living in dry climates, adaptations to minimize water loss include closing stomata, becoming dormant, altering leaf characteristics to minimize water loss, and losing leaves. Plant responses to short-term flooding include hormonal changes. Plant living in high salt concentration may exclude, secrete, or dilute salts that have been taken up. Halophytes are plant can take up water from saline soils by decreasing the water potential of their roots with high concentrations of organic molecules.

Review It

Mangroves are plants that live in soils that are continuously flooded by salt water. Describe how these plants are adapted to dealing with the challenges of their wet, salty environment.

37.6 Phloem Transport

Review It

Carbohydrates are manufactured in the leaves and green parts of the plant and distributed through the phloem in a process called translocation. Active transport of sugars into the phloem causes a reduction in water potential. Water then moves into the phloem and turgor pressure drives the carbohydrates to a sink, where the sugar is unloaded. Carbohydrates and their converted forms can be transported both up and down the plant. The substances that moves through phloem, called sap, can contain plant hormones, mRNA, and other substances in addition to sugars.

Essential Knowledge covered
2.B.2: Growth and dynamic homeostasis are maintained by the constant movement of molecules across membranes.

Review It

Determine if the following statements are true **(T)** or false **(F)**:

Transpiration is the process that describes the movement of carbohydrates in phloem within a plant.

Sap is a carbohydrate- and nutrient-rich fluid found in phloem.

Movement of carbohydrates in plants is unidirectional.

Phloem also transports plant hormones.

Use It

What is the pressure-flow model, and how does it relate to carbohydrate movement in a plant?

Summarize It

Compare and contrast how water and carbohydrates move within a plant.

Chapter 38: Plant Nutrition and Soil

Essential Knowledge

2.A.3	Organisms must exchange matter with the environment to grow, reproduce and maintain organization. **(38.3)**	Big Idea 2
2.D.3	Biological systems are affected by disruptions to their dynamic homeostasis. **(38.4)**	Big Idea 2
4.A.2	The structure and function of subcellular components, and their interactions, provide essential cellular processes. **(38.4)**	Big Idea 4
4.A.6	Interactions among living systems and with their environment result in the movement of matter and energy. **(38.4)**	Big Idea 4

Chapter Overview

Plants depend on the nutrients they can obtain from soils in order to grow. There are nine macronutrients and seven micronutrients that plants require for proper growth and development. Plants often have close associations with bacteria and fungi, and some have unique adaptations to capture and digest small animals, which allow plants to collect nutrients in ways other than soil extraction.

38.1 Soils: Substrates on Which Plants Depend

Recall It

Plants depend on the minerals, organic matter, water, air, and organisms that are found in the soil to grow. Roots become established in the mineral-rich layer of soil known as topsoil. Topsoil is formed from decaying organic matter. The microorganisms found in the soil are important for nutrient recycling. Most soil particles have negative charges that draw positively charged ions away from the roots. As plants require these ions, active transport is needed to draw mineral ions into the roots. Maintaining a correct balance of organic matter, water, and air in soil is important for plant growth. If the soil is too saline, there will be a loss of water and turgor in plants. The pH of the soil is also important as it affects mineral availability. It is important to practice cultivation techniques that reduce soil erosion and overuse of inorganic compounds such as fertilizers, pesticides, and herbicides.

Review It

List the components of topsoil.

Using the diagram provided, explain how soil particles can interfere with mineral uptake in plant roots.

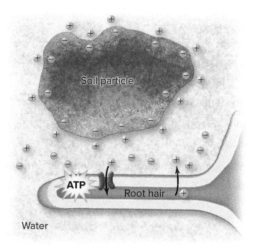

38.2 Plant Nutrients

Recall It

There are nine macronutrients required by plants for normal growth and development: carbon, oxygen, hydrogen, nitrogen, potassium, calcium, magnesium, phosphorus, and sulfur. The seven micronutrients required by plants are chlorine, iron, manganese, zinc, boron, copper, and molybdenum. Nutrient levels are of significant concern to humans we depend on crop productivity for our food sources. As a result, humans have spent much time on plant breeding efforts to increase nutrient levels in food crops.

Review It

Define macronutrients and micronutrients

Describe how you might determine if a particular element is essential for plant growth.

38.3 | Special Nutritional Strategies

Recall It

Plants need nitrogen in the form of ammonia or nitrate to build amino acids but are unable to use the most common atmospheric form, N_2. Therefore, many plants have evolved a symbiotic relationship with bacteria that can fix atmospheric nitrogen. In exchange for the nitrogen needed for protein synthesis, the plants provide carbohydrates to the bacteria. Plants more frequently have symbiotic relationships with mycorrhizal fungi, which allows the plants more access to the uptake of phosphorous and other nutrients by extending the surface area of the root system. Some plants have adapted to gain extra nutrients through digesting small animals, while others are parasitic and steal nutrients and carbohydrates from other plants.

Essential Knowledge covered
2.A.3: Organisms must exchange matter with the environment to grow, reproduce, and maintain organization.

Review It

Determine if the following statements on plant symbiosis are true or false **(T/F)**:

All plants have a symbiotic relationship with nitrogen-fixing bacteria.

Mycorrhizal fungi growing on plant roots provide the plants access to more micronutrients in exchange for carbohydrates.

Parasitic plants have a mutualistic symbiotic relationship with other plants.

Plants provide nitrogen-fixing bacteria with atmospheric N_2.

Use It

Describe a symbiotic relationship and a parasitic relationship that have evolved for plant species to obtain nutrients from other organism.

38.4 | Carbon-Nitrogen Balance and Global Change

Recall It

Increasing CO_2 levels can alter photosynthesis in plants. As CO_2 concentrations increase, the rate of photosynthesis increases and consequently biomass increases; however, the plant tissue that is produced is high in carbon relative to nitrogen, with a shift toward more carbohydrate and less protein. This leads to a decrease in the nutritional value of plants. Organisms must eat more plant matter to obtain the same amount of nutrients. Then, there is the loss of more plant matter through herbivory. Rising temperatures can affect plant respiration and carbon levels in plants and can cause additional changes in plant nutrient balance.

Essential Knowledge covered
2.D.3: Biological systems are affected by disruptions to their dynamic homeostasis.
4.A.2: The structure and function of subcellular components, and their interactions, provide essential cellular processes.
4.A.6: Interactions among living systems and with their environment result in the movement of matter and energy.

Review It

Recalling the process of photosynthesis from Chapter 8, what is the purpose of the Calvin cycle in plants?

The following diagram shows a possible outcome of increased CO_2 in the atmosphere and the effect on plant growth and human nutrient. Add arrows next to the blank statements to complete the statement.

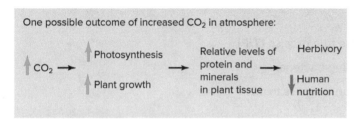

One possible outcome of increased CO_2 in atmosphere:

Use It

Would C_3 or C_4 plants be affected greater by the increase of CO_2 in the atmosphere? Explain your answer.

Why does respiration in plants change with increasing temperatures?

Recall It

Some plants have the ability to pick and accumulate high levels of contaminants in their roots and shoots. These plants are ideal for phytoremediation, the use of plants to concentrate or breakdown contaminants. Plants used for phytoremediation should be harvested and removed or else animals feed on these plants may be exposed to high concentrations of toxic compounds. Phytoremediation can be used to clean the soil, air, or water. Some plants can even metabolize organic compounds making them less toxic. Phytoremediation is an efficient and cost effective way to remove contaminates from large areas.

Review It

Using the information provided to you in this chapter describe how poplar trees have been used to remove trichloroethylene (TCE) from the soil.

38 Chapter Review

Summarize It

In the last 250 years, it is estimated that CO_2 levels have increased 46%, largely as a consequence of human activities. Rising CO_2 levels are correlated to increasing temperatures, as you may recall from Chapter 26. With these two pieces of evidence in mind, use what you know regarding photosynthesis and respiration to describe how a plant species today may be different from the same plant species 250 years ago.

Chapter 39: Plant Defense Responses

Extending Knowledge

Plants, like all living organisms, are affected by disruptions to their homeostasis. Homeostasis can be disrupted by environmental or biotic factors. Chapter 39 provides an overview of the defensive mechanisms plants use to protect and restore their homeostasis.

Chapter Overview

Plants have evolved many different physical and chemical defenses to protect themselves against predation and infection, as well as to enhance their competitive advantages. Physical defenses include thick waxy cuticles, and lipids like cutin and suberin. Some plants accumulate secondary metabolites that can poison herbivores, while others can produce toxic chemicals when leaves are wounded. Some plant species evolved symbiotic relationships with animals that physically protect the plant.

Review It

Determine if the following plant defense mechanism is chemical **(C)** or physical **(P)**.

A honey locus tree is covered in thorns.

Lima beans produce cyanide.

An above ground plant is covered in cutin.

Epidermal cells secrete wax.

An oak tree produces tannins.

A wounded leaf triggers the production of jasmonic acid.

Summarize It

Are all chemical plant defenses static?

Describe the types of systematic responses plants have to defend against invaders.

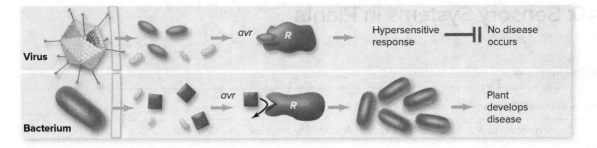

Explain why, in the diagram above, the bacteria can cause disease in the plant but the virus cannot. Assume both the virus and bacteria are both equally infectious.

Chapter 40: Sensory Systems in Plants

Essential Knowledge

2.C.2	Organisms respond to changes in their external environment. (40.1)	Big Idea 2
2.E.2	Timing and coordination of physiological events are regulated by multiple mechanisms. (40.1, 40.5)	Big Idea 2

Chapter Overview

Plants respond to light, gravity, mechanical stimuli, water, and temperature in a number of ways. Most changes are hard to observe at the macroscopic scale since changes occur at the cellular level. Hormones and other signaling molecules play a large role in how plants respond to their environment.

40.1 Responses to Light

Recall It

Many growth responses, such as seed germination and shoot elongation, are linked to phytochrome action. Phytochrome is a red-light receptor that exists as two interconvertible forms. The inactive form, P_r, absorbs red light and is converted to the active form, P_{fr}. P_{fr} absorbs far-red light and is converted back to the inactive form, P_r. P_{fr} enters the nucleus and binds with other proteins to form a transcription complex, leading to expression of light-regulated genes. P_{fr} can also activate a cascade of transcription factors. Far-red light inhibits germination, and red light stimulates it. Crowded plants receive a greater proportion of far-red light, which is reflected from neighboring plants. Crowded plants respond by growing taller to compete more effectively for sunlight. Phototropisms are directional growth responses of stems toward light. Positive phototropism bend toward blue light, possibly connected to blue light receptors called phototropins. Plants, like many other eukaryotes, appear to follow a circadian rhythm. Circadian rhythm refers to a cycle of activity in an organism during a 24-hour period. In the absence of light, the cycle's period may become desynchronized, but it resets when light is available. Circadian rhythms in plants are linked to phytochrome and blue light sensors.

Essential Knowledge covered
2.C.2: Organisms respond to changes in their external environment.
2.E.2: Timing and coordination of physiological events are regulated by multiple mechanisms.

Review It

Identify the light receptor involved in the process:

Structure	Process
	Stimulates growth responses in response to far-red light
	Stmulates growth response in response to blue light

Define circadian rhythm.

Use It

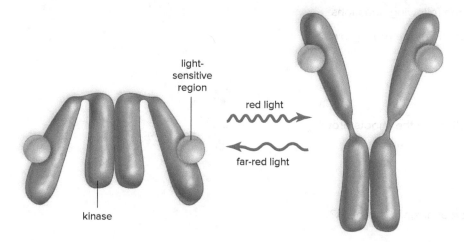

light-sensitive region

red light

far-red light

kinase

The figure above shows the conversion of phytochrome from the inactive form to the active form. Circle the form that will enter the nucleus and trigger the expression of light-regulated genes. Explain your choice.

Why will plants placed in the dark eventually become desynchronized in their processes?

40.2 Responses to Gravity

Recall It

Gravitropism is the response of a plant to the gravitational field of Earth. Shoots grow away from gravity, whereas roots grow towards gravity. These are known as positive and negative tropisms. The hormone auxin plays a primary role in gravitropic responses. Plastids that contain starch may also be involved in sensing gravity. Stems bend away from the center of gravity because of the presence of more auxin on the lower side in stems. Auxin causes more cells to grow in the lower side of stems than on the upper side.

Review It

Use the diagram on the right to answer the following questions:

In relationship to gravity, which direction are the roots growing?

In relationship to gravity, which direction are the shoots growing?

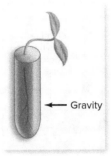

Gravity

Where is there a higher concentration of auxin in this plant?

40.3 Responses to Mechanical Stimuli

Recall It

Plants respond to touch-induced responses in a number of ways. Some changes are permanent, while others are temporary. Thigmotropism is a directional growth of a plant in response to mechanical stimuli. Thigmomorphogenesis is the permanent change in growth form in response to mechanical stress. Thigmonastic responses are independent of the direction of the stimulus and are usually produced by changes in turgor pressure. A stimulus causes an electrical signal, which results in a loss of potassium ions and water from cells. The loss of turgor causes the can cause leaves to move. Light can also induce changes in turgor pressure, resulting in leaf tracking of sunlight, flower opening, and leaf sleep movements.

Review It

Determine the process responsible for the following examples:

Example	Process
A vine grows around a trellis.	
Trees under constant exposure from northwest winds grow in the northwest direction.	
A venus fly trap closes around an insect.	

Recall It

Plants respond to changes in water and temperatures in different ways, depending on if the change is sudden or gradual. When exposed to rapid increases in temperature, plants can produce heat shock proteins, which help to stabilize other proteins. If temperatures suddenly fall, plants may increase unsaturated lipids in their membranes, limiting ice crystal formation to extracellular spaces, and produce antifreeze proteins. In response to more gradual weather changes, plants can go into dormancy. Dormancy is the cessation of growth that occurs when a plant is exposed to environmental stress. In deciduous trees in the fall, seasonal leaf abscission occurs. As discussed in Chapter 31, plant can also remain dormant in seed form, suspending germination until environmental conditions are optimal.

Review It

Determine whether the following statements are true or false **(T/F)** about plants responses to changes in temperature:

Seeds can enter long periods of dormancy.

All coniferous trees experience seasonal leaf abscission.

Some seeds can remain dormant for thousands of years.

Some plants produce antifreeze proteins to prevent ice crystals from forming when experiencing cold temperatures.

Plants can survive temperature extremes.

Heat-shock proteins are produced when a plant experiences a sudden drop in temperature.

40.5 Hormones and Sensory Systems

Recall It

Hormones are involved in many aspects of plant function and development. Hormones are produced in one part of a plant and then transported to another, where they bring about physiological or developmental responses. There are many different types of plant hormones. Auxins are produced in apical meristems and immature parts of a plant. Auxins can promote stem elongation, adventitious root formation, cell division, and lateral bud dormancy. Auxins can also inhibit leaf abscission and induce ethylene production. Cytokinins are produced in root apical meristems and immature fruits. Cytokinins can promote mitosis, chloroplast development, and bud formation. Cytokinins also delay leaf aging. Gibberellins are produced by root and shoot tips, young leaves and in seeds. Gibberellins promote the elongation of stems and the production of enzymes in germinating seeds. Ethylene induces fruit ripening and controls leaf, flower, and fruit abscission. It also suppresses stem and root elongation. Ethylene may also activate a defense response against pathogens and herbivores. Mature green leaves, fruits, root caps, and seeds produce abscisic acid (ABA). ABA suppresses growth and induces dormancy, inhibits the effects of other hormones, and controls stomatal closure.

Essential Knowledge covered
2.E.2: Timing and coordination of physiological events are regulated by multiple mechanisms.

Review It

Using your textbook, identify the plant hormone involved in the following plant process:

Process	Hormone
Stimulates seed production	
Induces ethylene production	
Promotes fruit ripening	
Induces dormancy	
Stimulates cell division	
Controls leaf abscission	

Use It

Imagine you are a grocery store owner, and you are shipped unripe tomatoes. How is it possible to ripen the tomatoes even though they are already off the vine?

Frits Went conducted an experiment that shed light on phototropisms in plants. Using the diagram below, explain how Frits Went concluded that the hormone auxin is what causes shoots to bend toward the light.

 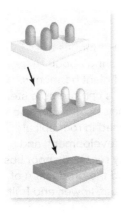  

At a cellular level, explain how the acid growth hypothesis explains how auxin causes changes in plant growth.

---

40 Chapter Review

Summarize It

The sensitive plant, *Mimosa pudica*, leaves fold when touched. Describe the internal mechanism that allows the plant to respond in such a way.

Chapter 41: Plant Reproduction

Essential Knowledge

2.A.1	All living systems require constant input of free energy. **(41.1)**	**Big Idea 2**
2.E.1	Timing and coordination of specific events are necessary for the normal development of an organism, and these events are regulated by a variety of mechanisms. **(41.5, 41.6)**	**Big Idea 2**
2.E.2	Timing and coordination of physiological events are regulated by multiple mechanisms. **(41.2)**	**Big Idea 2**
2.E.3	Timing and coordination of behavior are regulated by various mechanisms and are important in natural selection. **(41.2)**	**Big Idea 2**

Chapter Overview

Angiosperms have many different reproductive strategies. Just like animals, plants undergo developmental changes leading to reproductive maturity. Flowers and fruits that develop as a result of these developmental changes are often coordinated by environmental cues.

41.1 | Reproductive Development

Recall It

To begin flowering, a plant responds to external and internal signals. External factors include light and temperature. Internal factors include hormone production. The transition from reproductive growth to flowering in plants is termed phase change.

Essential Knowledge covered
2.A.1: All living systems require constant input of free energy.

Review It

Determine if the factor that triggers flowering is internal **(I)** or external **(E)**:

    Light

    Hormone production

    Temperature

Use It

How do external and internal factors work together to trigger growth and developmental changes in plants?

Recall It

Many factors can induce flowering in angiosperms. Light-dependent flowering is based not on the amount of light, but actually the length of the dark period a plant experiences during 24 hours. Light-dependent plants can be short-day (long period of dark), long-day (short period of dark), or day-neutral. Some plants require vernalization, or exposure of seeds or plants to chilling, in order to induce flowering. Some plants follow autonomous pathways to flower and do not depend on external cues. In these plants, balance between floral-promoting and floral-inhibiting signals controls flower development.

Essential Knowledge covered
2.E.2: Timing and coordination of physiological events are regulated by multiple mechanisms.
2.E.3: Timing and coordination of behavior are regulated by various mechanisms and are important in natural selection.

Review It

Identify what is required for following angiosperms to flower:

Plant Type	Requirements for Flowering
Short-day plant	
Day-neutral plant	
Long-day plant	

Define *vernalization*.

On what external cues do plants following an autonomous pathway depend in order to flower?

Use It

Suppose you pick up a packet of seeds to plant in your flower garden. On the back, it says plant the seeds 2–3 weeks before the last frost to ensure flowering. Explain the scientific reasoning behind these instructions.

A scientist is trying to classify a plant as a short-day or long-day plant. If the plant does not bloom when the night cycle is interrupted with a bright flash of light, what type of plant is it?

41.3 Structure and Evolution of Flowers

Recall It

Flowers are thought to have evolved from leaves. Flowers can be categorized as either complete or incomplete. Complete flowers have four whorls, which correspond to the four floral organs. Incomplete flowers lack at least one of these whorls. The four whorls are the calyx, corolla, androecium, and gynoecium. The androecium contains all the male structures of the flower, and the gynoecium contains all the female parts. The flower houses the haploid generations that produce gametes and increases the probability for successful gamete union.

Review It

Using your textbook, identify what flower parts make up the four whorls of a flower.

Whorl	Structure(s)
Calyx	
Corolla	
Androecium	
Gynoecium	

41.4 Pollination and Fertilization

Recall It

Pollination is the process by which pollen grains are placed on the stigma. There are many ways pollen can be carried to a flower. Some flowers are capable of self-pollination. Self-pollinating plants are usually found in stable environments and tend to produce uniform progeny. Other flowers depend on the wind or animals to carry pollen to the stigma. Wind pollination is a passive process and, as such, does not carry pollen over long distances. Wind-pollinated plants must grow relatively close together to each other and in dense groups. Animal pollinators provide an efficient transfer of pollen that may cover long distances. Flowers and animal pollinators have coevolved. Animal-pollinated flowers produce odors and visual cues to guide pollinators. Bees, birds, and bats are examples of animal pollinators.

Review It

Thinking in evolutionary terms, come up with some pros and cons for each of the different types of pollination in angiosperms.

Pollination Type	Pros	Cons
Wind pollination		
Self pollination		
Animal pollination		

41.5 Embryo Development

Recall It

During embryo development, some key structures are formed. First, a food supply is established for the embryo. In angiosperms, this is the endosperm; in gymnosperms, this is the megagametophyte. In addition, a seed coat is formed, and the fruit develops in the carpel. During embryogenesis, the root-shoot axis and the radial axis are formed. Three tissues are also formed: protoderm, ground meristem, and procambium. When these three differentiate, they become the three types of tissue in an adult plant.

Essential Knowledge covered
2.E.1: Timing and coordination of specific events are necessary for the normal development of an organism, and these events are regulated by a variety of mechanisms.

Review It

Identify if the statements are true **(T)** or false **(F)** about embryo development:

Angiosperm and gymnosperm have different food reserves for developing embryos.

The protoderm and procambium are the food for an embryo.

The endosperm differentiates to become adult tissue in a plant.

The root-shoot axis becomes established at an early stage during embryogenesis.

Use It

Why does the development of a food supply for a plant embryo an important step of embryogenesis?

41.6 Germination

Recall It

Just as in flowering, environmental signals are often needed for germination. Environmental cues may include light of a certain wavelength, an appropriate temperature, and stratification. Seed germination occurs when the first root emerges through the seed coat. In order to germinate, a seed must absorb water in order to break the seed coat. Abundant oxygen is then necessary for the high metabolic rate of newly emerged seedling. As germination is a high-energy process, stored nutrients such as starch, fats, and oils in the endosperm or embryo are quickly used up. To use the stored starch, the embryo produces the hormone gibberellic acid. Gibberellic acid, in turn, stimulates production of an amylase to break down amylose, a component of starch. Starch metabolism can be inhibited by abscisic acid, a hormone that has a role in dormancy. During seedling emergence in eudicots such as beans, the cotyledons are often pulled up with the growing shoot. In monocots such as corn, the cotyledon remains underground. A seedling enters the postembryonic phase of growth and development when the emerging shoot becomes photosynthetic.

Essential Knowledge covered
2.E.1: Timing and coordination of specific events are necessary for the normal development of an organism, and these events are regulated by a variety of mechanisms.

Review It

Create a flow diagram of the steps of seedling germination.

Use It

How does gibberellic acid modulate plant development during germination?

Recall It

Asexual plants reproduce through mitotic cell division, producing genetically identical individuals. There are different forms of asexual reproduction in plants. In apomixis, embryos are asexually produced from parent plants. A more common form is vegetative reproduction, where plants are cloned from parts of the adult plant. Forms of vegetative reproduction include runners, rhizomes, suckers, and adventitious plantlets. Plants can also be cloned in the laboratory from cells and tissues using tissue-culturing techniques.

Review It

Compare and contrast apomixis and vegetative reproduction.

Use your textbook to provide a description for each mode of vegetative reproduction:

Mode	Description
Runners	
Rhizomes	
Suckers	
Adventitious plantlets	

41.8 Plant Life Spans

Recall It

Plants have a variety of life spans. Life spans of plants are divided into three groups: perennials, annuals, and biennials. Perennials live for years, although they may undergo dormancy. Annual plants grow, reproduce, and die in a single year. Biennial plants follow a two-year life cycle. During the first year, biennials grow and store nutrients. In the second year, they produce flowers and seeds.

Review It

Use your textbook to provide at least one plant example for each of the different life spans:

Life Span	Example
Annuals	
Perennials	
Biennials	

Summarize It

1. How do light and temperature ensure that plants flower at optimal times?

2. A packet of seeds you planted in a garden did not yield any plants. What data would you collect in order to try to determine why the seeds did not germinate?

Chapter 42: The Animal Body and Principles of Regulation

Essential Knowledge

2.A.1	All living systems require constant input of free energy. **(42.8)**	Big Idea 2
2.A.2	Organisms capture and store free energy for use in biological processes. **(42.8)**	Big Idea 2
2.C.1	Organisms use feedback mechanisms to maintain their internal environments and respond to external environmental changes. **(42.7)**	Big Idea 2
2.D.2	Homeostatic mechanisms reflect both common ancestry and divergence due to adaptation in different environments. **(42.8)**	Big Idea 2
3.E.2	Animals have nervous systems that detect external and internal signals, transmit and integrate information, and produce responses. **(42.5)**	Big Idea 3
4.B.2	Cooperative interactions within organisms promote efficiency in the use of energy and matter. **(42.6)**	Big Idea 4

Chapter Overview

All vertebrates share the same basic body plan. Vertebrates have similar tissues and organs. This chapter explains the different levels of organization and describes how vertebrates maintain homeostasis.

42.1 Organization of the Vertebrate Body

Recall It

There are four levels of organization in a vertebrate body: cells, tissues, organs, and organs systems. Tissues are composed of groups of cells of a single type. The different types of vertebrate tissues are epithelial, connective, muscle, and nerve tissue. These tissues go on to form structural and functional units known as organs. Groups of organs that together form a function are known as an organ system.

Review It

Label the levels of vertebrate body organization in the illustration below:

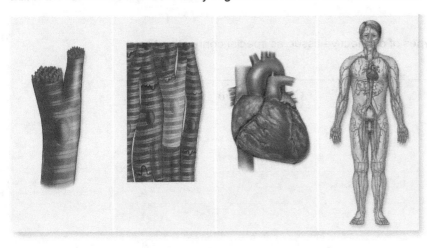

Recall It

Epithelium covers every surface of the vertebrate body, forming barriers and membranes. Epithelial cells are tightly bound together and have remarkable regenerative properties. There are two classes of epithelial membranes: simple and stratified. Simple epithelial membranes are one cell layer thick. Glands including the exocrine and endocrine glands are formed from invaginated epithelia. Stratified epithelial membranes are at least two layers thick and form the epidermis.

Review It

Using your textbook, describe different types of simple and stratified epithelial tissues.

Simple	Stratified

42.3 Connective Tissue

Recall It

Connective tissues are diverse in function. All share a common origin (the mesoderm) and all have abundant extracellular material. There are two major classes of connective tissue: connective tissue proper and special connective tissue. Connective tissues have fibroblasts but can differ in their ground substance. Specialized tissues have unique cells and matrices.

Review It

Use your textbook to classify the following types of connective tissue as special connective tissue or connective tissue proper:

Tissue	Classification
Loose connective tissue	
Cartilage	
Blood	
Dense connective tissue	
Bone	

42.4 | Muscle Tissue

Recall It

Muscle tissues contain cells with actin and myosin, making them specialized for contractions. There are three types of muscle tissue: smooth, skeletal, and cardiac. Smooth muscles power involuntary contractions, skeletal muscles powers all voluntary movement, and cardiac muscle powers the pumping of the heart.

Review It

Identify the following muscle tissue type as smooth, skeletal, or cardiac:

Muscle Tissue	Type
Found in the walls of the stomach	
Found in the walls of the heart	
Powers walking and talking	
Makes up the iris of the eye	

42.5 | Nerve Tissue

Recall It

Nerve tissue includes neurons and their supporting cells, neuroglia. Neurons have three parts: a cell body, dendrites, and an axon. The cell body has a nucleus, the dendrites receive impulses, and the axon transmits impulses away. Neuroglia help regulate the neuronal environment by insulating neurons with myelin sheaths and eliminating foreign materials in and around axons. The nervous system is composed of two major divisions. The central nervous system (CNS) is the brain and spinal cord, and the peripheral nervous system (PNS) contains nerves and ganglia.

Essential Knowledge covered
3.E.2: Animals have nervous systems that detect external and internal signals, transmit and integrate information, and produce responses.

Review It

Label the cell body, axon, and dendrites of the following axon:

Use It

How do nerve tissues allow animals to react to information or signals that occur outside the body?

42.6 Overview of Vertebrate Organ Systems

Recall It

Eleven major vertebrate organ systems will be covered in depth in the following chapters. Organ systems are grouped by the functions they provide. The integumentary system forms a barrier against attack. The immune system mounts a counterattack to foreign pathogens. The nervous and sensory systems are involved in communication and integration of external and internal stimuli. These two systems work in parallel with the endocrine system, which issues chemical signals in the body. The musculoskeletal system consists of muscles and the skeleton they act upon, allowing for movement and support. The digestive, circulatory, respiratory, and urinary systems accomplish ingestion of nutrients and elimination of wastes. Finally, the reproduction and development ensure continuity of species.

Essential Knowledge covered
4.B.2: Cooperative interactions within organisms promote efficiency in the use of energy and matter.

Review It

Name five different functions organ systems perform in the vertebrate body.

Use It

Describe an example of how multiple vertebrate systems work together to accomplish or regulate a bodily function.

Recall It

Homeostasis refers to the dynamic constancy of the internal environment and is essential for life. Negative feedback mechanisms are one way organisms maintain their internal environment. Negative feedback mechanisms occur where the end result of a process feeds back to limit the process. Negative feedback mechanisms keep values within a range. Negative feedback loops can be broken down into a stimulus, sensor, integrating center, effector, and response. Positive feedback mechanisms enhance change, as a change in one direction brings about further changes in the same direction.

Essential Knowledge covered
2.C.1: Organisms use feedback mechanisms to maintain their internal environments and respond to external environmental change.

Review It

Draw a picture of a negative feedback loop. Be sure to include the stimulus, sensor, integrating center, effector, and response. What role does each part play in the loop?

Use It

Using the example of the regulation of contractions during childbirth, describe what happens when a contraction occurs. What type of feedback mechanism is this?

Recall It

Thermoregulation, or regulation of body temperature, is an important aspect that all organisms have to deal with. As you will recall from earlier chapters, the enzymes that make life possible are affected by temperature. The simplest equation for modeling body temperature is body heat = heat produced + heat transferred. Heat can be transferred by radiation, conduction, convection, and evaporation. Organisms can be classified by how they maintain their body temperature. Ectotherms are more or less dependent on the environmental temperature regulate their internal temperature through behavior. Endotherms create their own heat through having a high metabolic rate and have behavioral and physiological adaptations for cooling off, such as sweating. Shivering thermogenesis occurs in endotherms to produce heat when conditions get too cold. Some endotherms also go into a state of dormancy called topor when it is too cold by reducing both metabolic rate and body temperature. Hibernation is an extreme form of torpor. The part of the brain that controls temperature change in mammals is the hypothalamus.

Essential Knowledge covered
2.A.1: All living systems require constant input of free energy.
2.A.2: Organisms capture and store free energy for use in biological processes.
2.D.2: Homeostatic mechanisms reflect both common ancestry and divergence due to adaptation in different environments.

Review It

Provide an example an endotherm and an ectotherm.

Identify the following forms of heat transfer:

Description	Heat Transfer
A direct transfer of kinetic energy between the molecules	
The transfer of heat from a hotter body to colder body	
The transfer of heat through the heat of vaporization	
The transfer of heat brought about by the movement of a gas or liquid	

Use It

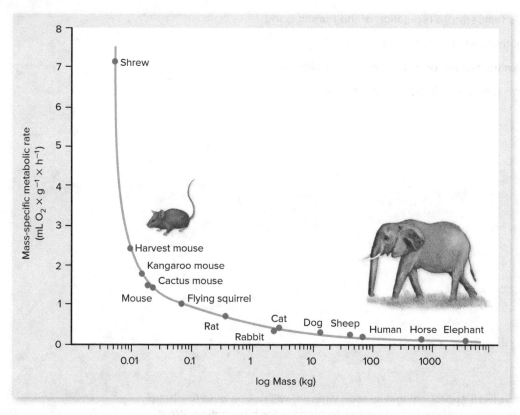

The diagram above illustrates the relationship between body mass and metabolic rate in mammals. Describe the relationship. What type of temperature regulation do the organisms in the diagram all share?

Compare and contrast the physiological responses of shivering and sweating in humans and their roles in thermoregulation.

Summarize It

1. The hypothalamus controls temperature regulation of the human body.

    a. How does the hypothalamus perform this regulation?

    b. What happens when the human body becomes too cold?

    c. What happens when the human body became too hot?

2. A midrange endotherm finds itself approaching a season where it will not be able to obtain as much food as it needs to keep itself warm and begins to prepare for hibernation. What physiological changes will occur to the endotherm as it enters this state?

Chapter 43: The Nervous System

Essential Knowledge

| 3.E.2 | Animals have nervous systems that detect external and internal signals, transmit and integrate information, and produce responses. (43.1, 43.2, 43.3, 43.4) | **Big Idea 3** |

Chapter Overview

In almost all animals, external and internal information is gathered by a network of nerve cells and integrated in some form of nervous system. The nervous system then issues commands to the body's muscles and glands in response to this information. The nervous system is a fast and efficient communication system, and helps maintain the body's homeostasis through feedback mechanisms.

43.1 Nervous System Organization

Recall It

As you learned in Chapter 42, the nervous system in humans is divided into the central nervous system (CNS) and the peripheral nervous system (PNS). These two systems work together to collect and integrate information, and carry out responses. Neurons are the cells which transmit nerve impulses. Neurons have a cell body, dendrites that receive information, and a long axon that conducts impulses away from the cell. There are three types of neurons in vertebrates: sensory, motor, and interneurons. Neuroglia are supporting cells of the nervous system, and include Schwann cells and oligodendrocytes. Schwann cells and oligodendrocytes produce myelin sheaths that surround and insulate axons. Nodes of Ranvier are small gaps that interrupt the myelin sheaths.

Essential Knowledge covered
3.E.2: Animals have nervous systems that detect external and internal signals, transmit and integrate information, and produce responses.

Review It

Determine if the following is a sensory **(S)**, motor **(M)** or interneuron **(I)**:

Carry impulses to muscles and glands

Provide links between sensory and motor neurons

Carry information about the environment to the CNS

Draw a picture of a typical vertebrate neuron. Include in your drawing: dendrites, a cell body, and a myelinated axon.

Use It

Explain the difference between a myelinated axon and an unmyelinated axon. What is myelin's function?

43.2 The Mechanism of Nerve Impulse Transmission

Recall It

Nerve impulses are transmitted through changes in electric potential across the neuron's plasma membrane. In neurons, high levels of K^+ are inside the cell, and high levels of Na^+ are outside of the cell. The resting potential (on average -70 mV), is a balance between diffusion of K^+ out of the cell and attraction back in by negative charge. Membrane potentials are changed when gated ion channels respond to chemical or electrical stimuli. The membrane may become depolarized (less negative) or hyperpolarized (more negative). Summation is the ability for these graded potentials to combine, amplifying or reducing their effects. A nerve impulse, or action potential, occurs when the depolarization level reaches around -55 mV. Action potentials are all-or-nothing events resulting from the rapid and sequential opening of voltage-gated ion channels and are propagated along axons. The influx of Na^+ during an action potential causes the adjacent region to depolarize, producing its own action potential. The velocity of nerve impulses increases as the diameter of the axon increases or if an axon is myelinated. Salutatory conduction, in which impulses jump from node to node, also increases speed.

Essential Knowledge covered

3.E.2: Animals have nervous systems that detect external and internal signals, transmit and integrate information, and produce responses.

Review It

In the following diagram, label the hyperpolarization, summation, and depolarization events.

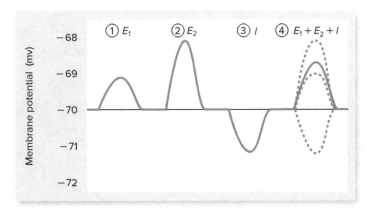

Use It

The neuron below is responding to a stimulus. Using what you have learned about how action potentials are formed, describe the process that is taking place below.

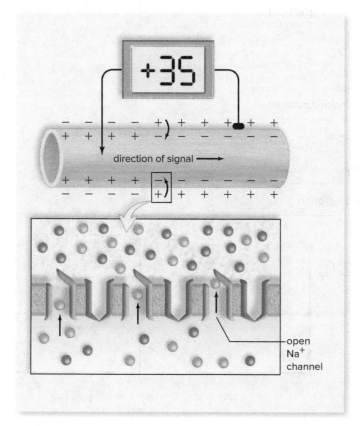

43.3 Synapses: Where Neurons Communicate with Other Cells

Recall It

An action potential cannot move by itself across a synapse (the gap between the axon and another cell). Synapses can be electrical or chemical. The majority of vertebrate synapses are chemical synapses, which release neurotransmitters to cross the synaptic cleft. There are many different types of neurotransmitter molecules, including acetylcholine, amino acids, biogenic amines (epinephrine, dopamine, serotonin), neuropeptides, and nitric oxide.

Essential Knowledge covered
3.E.2: Animals have nervous systems that detect external and internal signals, transmit and integrate information, and produce responses.

Review It

Identify the neurotransmitters based on the function it performs in the body:

Neurotransmitter	Function
	Muscle contractions
	Sleep and emotional states
	Body movement and brain functions
	"Fight or flight" responses
	Learning and memory
	Body movements and emotion

Use It

In the graphs to the right, which shows the effects of the neurotransmitter acetylcholine and which shows the effects of the neurotransmitter GABA? Explain your answer.

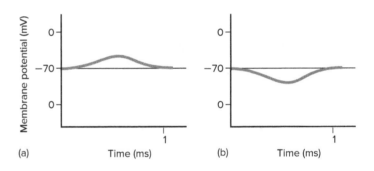

43.4 The Central Nervous System: Brain and Spinal Cord

Recall It

As animals became more complex, so did their nervous systems. The nervous system has evolved from a nerve net, to a nerve cord with associated nerves, to the complex coordination centers seen in many vertebrates. In vertebrates, the brain is divided into three sections: the hindbrain, the midbrain, and the forebrain. The hindbrain is composed of the cerebellum, pons, and medulla oblongata and is responsible for integration of and response to information from the muscles. The midbrain is composed of the optic tectum, which processes visual information. In mammals, the forebrain contains the cerebrum, which is the center for association and learning. The human forebrain exhibits exceptional information-processing ability. The cerebrum is divided into right and left hemispheres, which are subdivided into frontal, parietal, temporal, and occipital lobes. The spinal cord conveys messages and controls some responses directly. Reflexes are the sudden, involuntary movement of muscles in response to a stimulus.

Essential Knowledge covered
3.E.2: Animals have nervous systems that detect external and internal signals, transmit and integrate information, and produce responses.

Review It

Identify the structure of the central nervous systems based on the function provided:

Function	Structure
Processes reflexes involving the eyes and ears	
Processes information and houses learning	
Coordinates movements and balances	

List the organisms from least (1) complex nervous system to most (5) complex and describe the major characteristics of each.

Complexity	Organism	Nervous System
	Mollusk	
	Cnidarian	
	Earthworm	
	Crab	
	Human	

Use It

The relative size of the brain regions of vertebrates have changed as vertebrates evolved. Using the illustrations of the vertebrates below, describe the changes that have occurred leading up to the mammalian brain.

You may have heard of the expression "a knee-jerk reaction." Knowing what you know now about the nervous system, explain the physiology behind a knee-jerk reaction.

Recall It

The peripheral nervous system (PNS) is responsible for receiving information from the environment and conveying it to the central nervous system (CNS). The PNS is divided into the somatic system and the automatic system. Sensory receptors generate nerve impulses when simulated which brings about a response to the stimulus from an effector. The autonomic system is divided into the parasympathetic and sympathetic motor pathways which are respectively thought of as the "fight or flight" and "rest and digest."

Review It

Identify if the following effects are caused by the somatic or autonomic systems:

Stimulate muscles to contract

Supplies involuntary effectors

Fight of flight response

Determine if the following statements regarding the PNS are true or false **(T/F)**:

The PNS has somatic and autonomic systems.

The cerebrum is part of the PNS.

The parasympathetic division nerve effects include increasing heart rate and decreasing activity of the digestive organs.

Norepinephrine released in the sympathetic pathway prepares the body for action.

The PNS contains sensory neurons which send information to the CNS.

43 Chapter Review

Summarize It

Describe basic organization of the vertebrate brain and explain why they differ in proportions in different vertebrates.

Chapter 44: Sensory Systems

Extending Knowledge

As you learned in Chapter 43, animals have nervous systems that allow them to detect and respond to both internal and external signals. Chapter 44 provides additional details and examples of sensory systems, which animals evolved to receive and respond to specific stimuli.

Chapter Overview

Sensory systems in vertebrates include exteroceptors – sensory receptors that provide information from the external environment, as well as interoceptors – sensory receptors that receive information about internal conditions. Receptors are also classified by the type of signal they receive. Mechanoreceptors respond to pressure and vibration. Animals use them in touch, hearing, and balance. Chemoreceptors are used in smell and taste. Electromagnetic receptors react to heat and light, and include receptors such as the photoreceptors of the eye. As described in Chapter 43, sensory neurons pick up sensory information and send it to the central nervous system (CNS) where the information is processed (and responses originate).

Review It

Identify what type of receptor is used in detecting the following signals: mechanical **(M)**, chemoreceptor **(C)**, or electromagnetic **(E)**.

Sound

Light

Smell

Taste

Vibration

Summarize It

The ability for a human to hear is controlled by mechanoreceptors. Using your textbook, explain briefly how these receptors work and how humans receive sound information.

Compare and contrast rods and cones of the human eye.

Chapter 45: The Endocrine System

Essential Knowledge

3.D.2	Cells communicate with each other through direct contact with other cells or from a distance via chemical signaling. **(45.1, 45.3)**	**Big Idea 3**
3.D.3	Signal transduction pathways link signal reception with cellular response. **(45.2)**	**Big Idea 3**

Chapter Overview

The nervous and endocrine systems work together to regulate the body's processes. The endocrine system contains many specialized cells and glands, which produce a variety of chemical messengers and hormones. These messengers control many important processes, from growth and development to digestion and reproduction.

45.1 Regulation of Body Processes by Chemical Messengers

Recall It

A hormone is a signaling molecule carried by the blood that affects a distant target. The endocrine system produces hormones through specialized organs. The three classes of endocrine hormones are: (1) peptides and proteins, (2) amino acid derivatives, and (3) steroids. Hormones can also be categorized as lipophilic or hydrophilic. Lipophilic hormones are fat-soluble and can cross the cell membrane. Hydrophilic hormones, on the other hand, are water-soluble and cannot cross membranes. Paracrine regulators act in the organ where they are produced. Pheromones are chemicals released into the environment and used by individuals of the same species to communicate.

Essential Knowledge covered
3.D.2: Cells communicate with each other through direct contact with other cells or from a distance via chemical signaling.

Review It

Using your textbook, explain the transmission patterns of the chemical messengers below.

Hormones carried by blood	Paracrine secretions	Autocrine signals in cancer cells	Pheromones

Use It

Growth factors are important paracrine regulators. What are growth factors and how do they work?

45.2 Actions of Lipophilic Versus Hydrophilic Hormones

Recall It

The two classes of hormones, lipophilic and hydrophilic, have different receptors and actions. Circulating lipophilic hormones are carried in the blood, bound to transport proteins. These hormones are able to pass through the plasma membrane and activate intracellular receptors. The hormone-receptor complex can bind to specific gene promoter regions called hormone response elements to activate transcription. Hydrophilic hormones cannot cross the plasma membrane, so they work in a different way by activating receptors on target cell membranes. Once hydrophilic hormones bind to membrane receptors, they may initiate a signal transduction pathway through activating protein kinases. As you learned in Chapter 9, receptor tyrosine kinases autophosphorylate, initiating signal transmission. Other hydrophilic hormones use G protein–coupled receptors to activate effector proteins that produce second messengers.

Essential Knowledge covered
3.D.3: Signal transduction pathway link signal reception with cellular response.

Review It

Describe the major difference between lipophilic and hydrophilic hormones.

Define *receptor tyrosine kinases* and *G-protein coupled receptors* (review Chapter 9 if you're unsure):

Use It

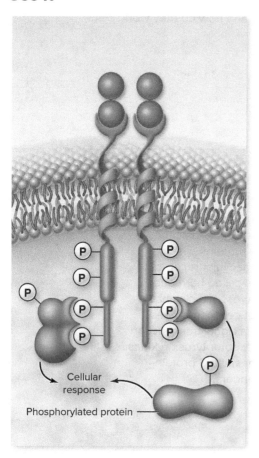

The diagram to the left shows the effect of a hormone on a cell. What type of hormone is present in the diagram? Describe the pathway the hormone is initiating.

45.3 The Pituitary and Hypothalamus: The Body's Control Center

Recall It

The pituitary gland and hypothalamus are endocrine glands that work together to regulate water reabsorption, growth, and the activity of other endocrine glands. The antidiuretic hormone (ADH) is a neurohormone released from the posterior pituitary that stimulates water reabsorption in the kidneys. The posterior pituitary also secretes oxytocin, which is needed during childbirth and in breast-feeding. There are many hormones produced by the anterior pituitary that stimulate growth and the expression of other endocrine glands. The activity of the anterior pituitary is regulated through the release and inhibition of hormones produced in the hypothalamus and regulated by negative feedback. A classic example is the production of thyroid stimulating hormone. Thyroxine is a chemical produced by the thyroid in response to thyroid stimulating hormone (TSH). Thyroxine inhibits further secretion of TSH.

Essential Knowledge covered
3.D.2: Cells communicate with each other through direct contact with other cells or from a distance via chemical signaling.

Review It

List two areas the thyroid regulates.

Use It

Your textbook relates the incredible tale of a man who was shot in the head and developed an urge to urinate every 30 minutes. A bullet had been lodged in his posterior pituitary, disrupting normal function. Which hormone production was affected, and how did this lead to his condition?

The following diagram details the regulation of the thyroid gland. Describe the process.

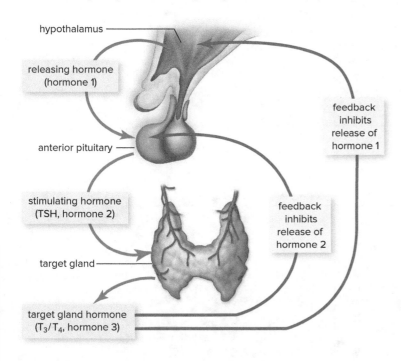

Recall It

While some endocrine glands are controlled by tropic hormones of the pituitary, others are independent of pituitary control. The thyroid gland, parathyroid gland, adrenal glands, and the pancreas also play important roles in regulating homeostasis. The thyroid and the parathyroid glands produce hormones that are necessary for nutrient uptake and mineral metabolism. The hormones thyroxine and triiodothyronine regulate basal metabolism in vertebrates. These hormones are also used to trigger metamorphosis in amphibians. Blood calcium is regulated by calcitonin and parathyroid hormone, as well as vitamin D. Corticosteroids maintain glucose homeostasis and modulate some aspects of the immune response. The pancreas secretes insulin, which reduces blood glucose, and glucagon, which raises blood glucose.

Review It

Describe the function of the hormone and identify the gland or organ it comes from:

Hormone	Function	Gland/Organ
Parathyroid hormone		
Aldosterone		
Cortisol		
Glucagon		
Insulin		
Calcitonin		
Thyroxine		

Recall It

Sexual development and reproduction are regulated by sex steroids. The ovaries primarily produce estrogen and progesterone, which are responsible for the menstrual cycle. The testes produce testosterone. Melatonin is produced by the pineal gland, and is used in the dispersion of pigment granules, as well as daily sleep–wake cycles. The thymus secretes hormones that regulate the immune system. The right atrium of the heart secretes atrial natriuretic hormone, which acts antagonistically to aldosterone. The skin manufactures and secretes vitamin D. Insect hormones control molting and metamorphosis. In insects, the hormone ecdysone stimulates molting, and juvenile hormone levels control the nature of the molt.

Review It

Determine if the following statements concerning hormones are true or false **(T/F)**:

The testes and the ovaries are important endocrine glands.

Atrial natriuretic hormone plays an important role in the regulation of sleep cycles.

Melatonin levels in your blood increase in the fall during the daytime.

Hormones play a role in the metamorphosis of a caterpillar into a butterfly.

All hormones are produced by endocrine glands.

Describe two ways eukaryotic cancer may alter the endocrine system.

Summarize It

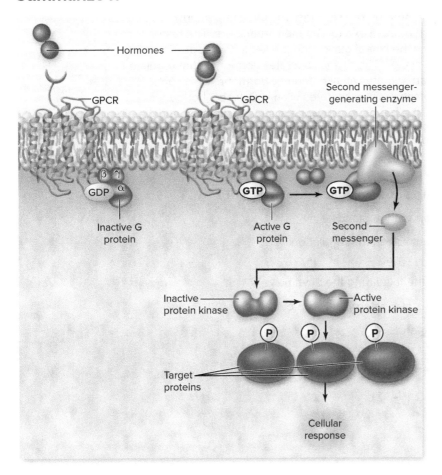

The diagram above shows a common but powerful response to a hormone expressed in the human body. Describe how this hormone is able to elicit a cellular response.

Chapter 46: The Musculoskeletal System

Extending Knowledge

As you have learned, animals exhibit complex properties that arise from interactions between different coordinating systems. Muscle contractions are an example of how the musculoskeletal and nervous systems act together to allow animals to move and perform various essential functions.

Chapter Overview

Muscles and skeletons are necessary for locomotion in vertebrates. There are three types of skeletons: hydroskeletons, exoskeletons, and endoskeletons. The bones of the musculoskeletal system are classified by how they develop. In humans, bones contain bone cells, blood capillaries, and nerves. There are different types of joints that are found in the musculoskeletal system, as well as skeletal muscles, which allow for different ranges of motions. Muscle fibers move as the result of the overlap of actin and myosin filaments. The two major types of skeletal muscle fibers are slow-twitch fibers (endurance) and fast-twitch fibers (power bursts). Twitches are the intervals between contraction and relaxation of a single muscle stimulation. Different modes of animal locomotion present different environmental difficulties, such as friction with water or the pull of gravity, that have caused unique adaptations to evolve in the vertebrate world.

Review It

Determine if the following statement pertains to a hydroskeleton **(H)**, exoskeleton **(EX)**, or endoskeleton **(EN)**:

Ants are supported by an external covering composed of chitin

A fluid-filled cavity allows earthworms to wriggle along the ground.

Fishes have rigid internal structure composed of connective tissues.

Draw a diagram of how actin and myosin interact. It may be helpful for you to review Chapter 4.

Summarize It

Describe difference between the movement of your eyes as you read this sentence and the movement in someone's legs as they run in terms of motor units.

How is a muscle stimulated to contract by a motor neuron? Be sure to identify the neurotransmitter, and the role of calcium and sodium in your answer.

Chapter 47: The Digestive System

Extending Knowledge

As you have learned, animals exhibit complex properties that arise from interactions between different coordinating systems. Chapter 47 illustrates this property through the exploration of animal digestive systems. Animal digestive systems are composed of many components from the cellular to organ level which coordinate together to allow animals to digest food and eliminate wastes.

Big Idea 2

Big Idea 4

Chapter Overview

The digestive tract is one of the major systems of the vertebrate body plan. Like the nervous system, the digestive system has evolved specialized and complex features in modern vertebrates. Invertebrates such as cnidarians and flatworms have very simple, sac-like guts. Specialized regions of the digestive tract make fragmentation, digestion, and absorption of food more efficient. The digestive system of humans has many specialized compartments and accessory organs. Teeth and salivary glands are used for initial processing of food. The esophagus moves food to the stomach where it is broken down in acidic juice, producing chyme. From there, nutrients are absorbed by the vast surface area of the small intestine, covered in villi. Many accessory organs, including the pancreas, liver, and gallbladder, aid in the digestive process. The pancreas and liver also regulate blood glucose concentrations.

Review It

List the organisms from least complex digestive system (1) to most complex (4) and describe the major characteristics of each.

Complexity	Organism	Digestive System
	Nematode	
	Earthworm	
	Cnidarians	
	Human	

List three accessory organs of the human digestive system.

Summarize It

Describe the major evolutionary steps of the digestive system using the hydra, nematode, earthworm, and salamander as your examples.

Describe the function of the small intestine, large intestine, and liver in human digestion.

Explain the role of villi and microvilli in the small intestine.

Use the following diagram to explain how the accessory organs, the liver and the pancreas, work together to regulate blood glucose levels.

High Blood Glucose Levels

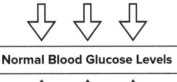

Normal Blood Glucose Levels

Low Blood Glucose Levels

Chapter 48: The Respiratory System

Extending Knowledge

As you have learned, animals exhibit complex properties that arise from interactions between different coordinating systems. Chapter 48 illustrates this through the exploration of animal respiratory systems. Animal respiratory systems are composed of many components from the cellular to organ level, which coordinate together to allow animals to exchange gases.

Chapter Overview

The respiratory and circulatory systems directly support all other vertebrate organ systems through facilitating gas exchange and transport of materials. All multicellular animals need to obtain oxygen and dispose of excess carbon dioxide. Surface area for gas exchange is important, as is the gradient of gases. Gills are one type of respiratory structure, and fishes have evolved an effective system of countercurrent flow, which increases their efficiency. Cutaneous respiration allows for the exchange of gases across the skin. Tracheal systems are networks of air ducts found in arthropods. Lungs evolved in terrestrial vertebrates to allow gas exchange while minimizing evaporation. Terrestrial vertebrates breathe in different ways, but all breathing takes advantage of the partial pressure of gases. In mammals, normal breathing rates keep the levels of carbon dioxide and oxygen at a specific range. Vertebrates rely on hemoglobin, an oxygen-carrying molecule. Oxygen combines with hemoglobin in the lungs. The oxygenated-blood is then carried to other cells in the body.

Review It

Describe the respiratory systems found in the following animals:

Animal	Respiratory System
Bird	
Fish	
Human	
Insect	
Amphibian	

Summarize It

Fish use a countercurrent exchange system to respire, as illustrated in the diagram below. How does this process allow for efficient blood oxygenation?

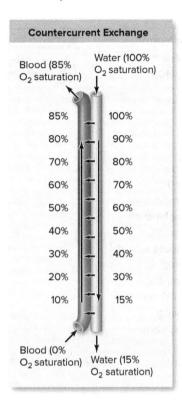

Countercurrent Exchange

Blood (85% O₂ saturation)   Water (100% O₂ saturation)

85%	100%
80%	90%
70%	80%
60%	70%
50%	60%
40%	50%
30%	40%
20%	30%
10%	15%

Blood (0% O₂ saturation)   Water (15% O₂ saturation)

Compare and contrast how amphibians and reptiles breathe.

Chapter 49: The Circulatory System

Extending Knowledge

As you have learned, maintaining homeostasis is important for living organisms, and regulatory mechanisms can be used to trace common ancestry as well as divergence. Chapter 49 illustrates this through the exploration of animal circulatory systems, which are involved in the transport of materials and the maintenance of internal conditions.

Chapter Overview

As you will recall from Chapter 42, the circulatory system is responsible for transporting nutrients and dissolved gases through the body. Blood is the primary circulatory fluid in vertebrates, composed of plasma and different cell types. Blood is involved in transporting materials, regulating body temperature and processes, as well as protecting the body. Invertebrate circulatory systems are less complex. Sponges, hydra, and nematodes do not have a true circulatory system; water circulates materials in and out of the organisms. Other invertebrates have either an open or a closed circulatory system. All multicellular organisms with a circulatory system have a heart to plump fluid through the body. Fishes have a two-chambered heart, amphibians and most reptiles have a three-chambered heart, and crocodiles, birds, and mammals have a four-chambered heart. In four-chambered hearts, the right side of the heart receives deoxygenated blood and sends it to the lungs, and the left side of the heart receives oxygenated blood and sends it to the body.

Review It

Determine if the following statement is true or false **(T/F)**:

Blood cells arise from stem cells.

Open circulatory systems move fluids in a loop.

Closed circulatory systems move fluids in a one-way path.

Grasshoppers have hearts.

Circulation in the hydra occurs in the gastrovascular cavity.

Describe two ways that the circulatory system protects the body.

Draw a diagram of a four-chambered heart. Show the movement of oxygenated and deoxygenated blood in your drawing.

Summarize It

Explain the differences between an open circulatory system found in some invertebrates and a closed circulatory system found in vertebrates.

Despite being distantly related through evolutionary history, why do birds and mammals share an extremely similar structured four-chambered heart?

Chapter 50: Osmotic Regulation and the Urinary System

Extending Knowledge

Animals have multiple organ systems which work together to perform complex tasks and maintain homeostasis. Chapter 50 illustrates this through the exploration of animal urinary and osmotic regulation systems, which are involved in the removal of wastes and the maintenance of internal conditions.

Chapter Overview

Vertebrates produce many types of nitrogenous waste through the catabolism of amino acids and nucleic acids. Depending on the type of nitrogenous waste produced, different forms of waste removal are necessary. Ammonia, urea, and uric acid are three forms of nitrogenous waste that are excreted by vertebrates. Different systems of waste removal include tubules in invertebrates and kidneys in vertebrates. Urinary systems have evolved in vertebrates to help them retain water and maintain the necessary ion levels present in their bodies. The environment a vertebrate lives in drives which wastes they produce and how they eliminate it.

Review It

Identify the type of nitrogenous waste excreted by the following animals. Place a star next to the waste that is eliminated directly:

Animals	Waste
Mammals, amphibians, and cartilaginous fishes	
Reptiles, birds, and insects	
Bony fish and aquatic invertebrates	

Identify the organ that produces vertebrate urine by filtration, secretion, and selective reabsorption of water and solutes.

Describe the difference between hypertonic, isotonic, and hypotonic solutions (hint: think back to Chapter 5).

Summarize It

Which vertebrates are able to produce urine with a higher osmotic concentration than their body fluids? How is this possible?

Which produces more urine, a freshwater fish or a marine fish, and why?

Chapter 51: The Immune System

Essential Knowledge

2.D.4	Plants and animals have a variety of chemical defenses against infections that affect dynamic homeostasis. **(51.1, 51.2, 51.3, 51.4)**	**Big Idea 2**
3.C.3	Viral replication results in genetic variation, and viral infection can introduce genetic variation into hosts. **(51.7)**	**Big Idea 3**

Chapter Overview

Microorganism and viruses constantly deliver threats of infectious disease, and the immune system has evolved to help defend us from these invasions. The immune system can be spilt into innate and adaptive immunities. Both are important in recognizing invading pathogens and bringing about a defense response.

51.1 Innate Immunity

Recall It

Pathogens can enter the body through the skin, the digestive tract, the respiratory tract, and the urogenital tract. These systems all have some sort of innate defense mechanism to deal with invading pathogens. The innate immune system identifies specific molecules of particular pathogens and brings a rapid response to sites of infection. The skin has a low surface pH, lysozyme secreted in sweat, and a population of nonpathogenic organisms, all of which deter pathogens. The epithelia of the digestive, respiratory, and urogenital tracts produce mucus to trap microorganisms. Innate receptor proteins include toll-like receptors, which have leucine-rich regions that recognize molecules such as LPS and peptidoglycan. The binding of these innate receptors to pathogen-specific molecules stimulates the production of antimicrobial peptides and the activation of complement. The body also has cells that specifically kill invading pathogens. These are different types of leukocytes: macrophages, neutrophils, and natural killer cells.

Essential Knowledge covered
2.D.4: Plants and animals have a variety of chemical defenses against infections that affect dynamic homeostasis.

Review It

List three possible routes of entry a pathogen may take to enter the body.

Describe the function of the three types of leukocytes listed below:

Leukocyte	Function
Macrophage	
Neutrophil	
Natural killer	

Use It

Describe an example of a nonspecific and a specific method vertebrates use to protect themselves from invading pathogens.

How are natural killer cells different from macrophages?

51.2 Adaptive Immunity

Recall It

Adaptive or acquired immunity involves genetic rearrangements to attack specific pathogens. When an individual or population is exposed to a particular pathogen, the surviving individuals can become immune to it. Antigens are molecules that provoke specific immune responses. Surface receptors on lymphocytes recognize antigens and direct a specific immune response. B lymphocytes, or B cells, secrete proteins called immunoglobulins (Ig) or antibodies when responding to antigens, taking part in humoral immunity. T lymphocytes, or T cells, regulate the immune response of other cells or directly attack cells as part of cell-mediated immunity. Adaptive immunity can be passive or active.

Essential Knowledge covered
2.D.4: Plants and animals have a variety of chemical defenses against infections that affect dynamic homeostasis.

Review It

Determine if the following statement applies to either **B** or **T** cells:

Secretes immunoglobulins

Initiates a signal pathway that leads to the production of
plasma cells that secrete antibodies

Participates in cell-mediated immunity

Does not secrete antibodies

Use It

Describe the adaptive immune system. How does it respond to a pathogen that has invaded your
body once before?

Recall It

T cells are classified as cytotoxic T cells (T_c) or helper T cells (T_H). T_c, also known as killer T cells,
eliminate virally infected cells and tumor cells. They destroy cells in a fashion similar to natural
killer cells by inducing apoptosis in infected or tumor cells. T_H cells secrete proteins that direct
immune responses. T_H cells promote both cell-mediated and humoral immune responses. Most
cells have glycoproteins encoded by the major histocompatibility complex (MHC) on their surfaces.
To be activated, T cells recognize foreign peptide fragments bound to MHC proteins. There are
two different classes of MHC proteins. MHC class I proteins are found on every nucleated cell and
proteins bound to these induce responses from T_c cells. MHC class II proteins are found only on
antigen-presenting cells, and T_H cells respond to peptides bound to this class.

Essential Knowledge covered
2.D.4: Plants and animals have a variety of chemical defenses against infections that affect dynamic homeostasis.

Review It

Fill in the chart to describe how lymphocytes recognize targets:

Lymphocyte	Class of MHC proteins recognized	Cells types that carry recognized MHC
Helper T cells (T_H)		
Cyotoxic T cells (T_c)		

Use It

How are T_H cell and T_C cells different?

51.4 Humoral Immunity and Antibody Production

Recall It

B lymphocytes, or B cells, secrete proteins called immunoglobulins (Ig) or antibodies when responding to antigens, taking part in humoral immunity. B cells are activated by membrane Ig molecules binding to a specific spot, called an epitope, on an antigen. Activated B cells produce antibody-secreting plasma cells and memory cells. Igs are Y-shaped molecules, with two light-chain and two longer heavy-chain polypeptides. The binding site is on one of the arms and is called the Fab region. Ig can agglutinate, precipitate, or neutralize antigens. Ig diversity is generated through DNA rearrangement. The secondary response to an antigen is more effective than the primary response. On second exposure to a pathogen, a rapid secondary immune response is launched due to memory cells.

Essential Knowledge covered
2.D.4: Plants and animals have a variety of chemical defenses against infections that affect dynamic homeostasis.

Review It

Draw a diagram of an immunoglobulin molecule. Be sure to denote where antigens bind.

Use It

If you have had chickenpox or been vaccinated against the disease, it is unlikely that you will contract it if exposed to the virus. How do B cells play a role in this phenomenon?

51.5 Autoimmunity and Hypersensitivity

Recall It

Occasionally, the immune system will attack the body's own tissues. This results in autoimmune diseases. Autoimmune diseases occur as the result of a loss of immunological tolerance. T cells and B cells act against the body's own cells, causing inflammation and organ damage. Allergies occur when the body produces a greater response to an antigen that is actually required.

Review It

Compare and contrast autoimmune diseases and allergies.

51.6 Antibodies in Medical Treatment and Diagnosis

Recall It

Human blood types, ABO, are determined by antigens present on the red blood cell surfaces. The immune system is tolerant to its own red blood cell antigens but it will create antibodies that bind to those that are different. Rh antigens are another type of blood-borne antigen that is either present or absent. Rh-negative antigens will produce antibodies against Rh-positive blood. For these reasons, medical procedures, such as blood transfusions, require blood typing of an individual.

Review It

List three antigens that could be found on a red blood cell.

A mother that is pregnant with a child has a blood type of AB⁻ (Rh negative). The father of the child has a blood type of O⁺. What possible complication may arise if the fetus has Rh⁺ blood?

51.7 Pathogens That Evade the Immune System

Recall It

Pathogens that become established and produce infection in a host have evaded nonspecific and specific immune responses. Some pathogens, such as the influenza virus, have surface antigens that change frequently, avoiding immune system recognition. Some bacteria have evolved mechanisms to inhibit normal immune system processes, such as slowing phagocytosis. HIV infection attacks the adaptive immune system directly by destroying T_H cells. This leads to immunosuppression; the body's ability to fend off infections, leading to the characteristic illnesses of AIDS.

Essential Knowledge covered
3.C.3: Viral replication results in genetic variation, and viral infection can introduce genetic variation to the hosts.

Review It

List three ways pathogens can evade or attack the immune system.

Use It

Explain why individuals that contract HIV usually die because of another disease such as cancer or pneumonia.

Summarize It

1. In a famous set of experiments that started in 1796, an English doctor named Edward Jenner documented the immune response to the disease cowpox. He inoculated individuals with cowpox and found they were immune to it when exposed to the disease again. We know now today that these responses are due to the immune system

    a. Use the graph below to describe what happened in the bodies of the individuals in the experiments.

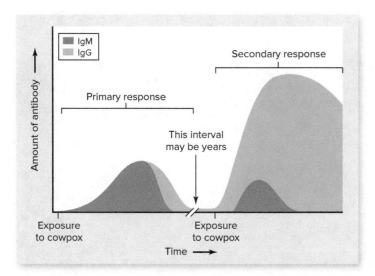

    b. Jenner later discovered that the individuals in his study also had an immunity to smallpox. While cowpox and smallpox are different viruses, what can you infer is similar the structure of the two viruses from Jenner's findings?

2. Using what you have learned about the virus that causes AIDS, construct an explanation on why AIDS is considered one of the most serious diseases in human history.

Chapter 52: The Reproductive System

Extending Knowledge

As you learned in Chapter 16, gene expression is controlled by numerous intra- and intercellular factors. Chapter 52 illustrates this through the exploration of animal reproductive systems, which consist of many coordinating organs and transduction pathways which work together to produce viable and diverse offspring.

Big Idea 3

Chapter Overview

Reproduction can occur asexually or sexually. Asexual reproduction produces offspring with the same genes as the parent organism, and can occur through budding or parthenogenesis. As describe in Chapter 11, sexual reproduction involves production by meiosis of haploid gametes (eggs and sperm). In some species, individuals can be both male and female. In hermaphroditism, an individual has both testes and ovaries, either simultaneously or sequentially. In some animals, the temperature an individual experiences as an embryo determines its sex. In mammals, sex is genetically determined by the presence of a Y chromosome. The Y chromosome carries the SRY gene. If this gene is present, the gonads of an embryo are converted into testes and the offspring will be male. In humans, female infertility ranges from failure of oocyte production to failure of zygote implantation. Male infertility is usually due to reduction in sperm number, viability, or motility; hormonal imbalance; or damage to the sperm delivery system. Hormonal treatment may be used to correct ovulatory defects or sperm production defects, and is used as an effective form of birth control in females. Birth control pills contain analogs of progesterone and estrogen that act by negative feedback to inhibit ovulation.

Review It

Describe the following reproductive strategies:

Reproductive Strategy	Description
Budding	
Hermaphroditism	
Parthenogenesis	
Sexual reproduction	

List two different ways sex may be determined during development in different sexually reproducing species.

Human females of reproductive age experience cycles of follicle-stimulating hormone (FSH) and luteinizing hormones (LH) which allows for egg development. What other hormones do FSH and LH regulate in the female body?

Summarize It

Describe the pathway that birth control pills alter and the effect on ovulation and reproduction.

Chapter 53: Animal Development

Extending Knowledge

As you learned in Part III, the production of offspring depends on the timing and coordination of developmental events, as well as the successful transfer of heritable information through meiosis and fertilization. Chapter 53 illustrates this through the exploration of animal development, which consists of many coordinating processes from the cellular to the organismal level.

Big Idea 2

Big Idea 3

Chapter Overview

Animal development is broken into four stages: fertilization, cleavage, gastrulation, and organogenesis. As you learned in Chapter 11, eggs and sperm are haploid (*n*) gametes. Eggs and sperm fuse during fertilization to result in a diploid (2*n*) zygote. During cleavage, the zygote rapidly divides into many cells and ends with the formation of a blastula. During gastrulation, three primary germ layers form. These germ layers interact in various ways during organogenesis to give rise to the organs of the body. Cell-signaling molecules and transcription factors control organogenesis.

Review It

Determine if the following statements concerning animal development are true or false **(T/F)**:

The ectoderm, mesoderm, and endoderm are three primary germ layers.

The haploid gametes, egg and sperm, are produced through mitosis.

Fertilization is the first step in animal development.

Changes in gene expression lead to organogenesis.

During cleavage, three germ layers form.

Draw a diagram of how meiosis leads to the production of haploid gametes. It may be helpful for you to review Chapter 11.

Summarize It

Describe how the process of fertilization produces a diploid zygote.

Using the information provided to you in Chapter 53.4, identify the molecular components and events necessary for the development of the heart in *Drosophila melanogaster*. How is this related to heart development in humans?

Chapter 54: Behavioral Biology

Essential Knowledge

2.E.3	Timing and coordination of behavior are regulated by various mechanisms and are important in natural selection. **(54.3, 54.4, 54.5)**	**Big Idea 2**
3.E.1	Individuals can act on information and communicate it to others. **(54.1)**	**Big Idea 3**
4.B.3	Interactions between and within populations influence patterns of species distribution and abundance. **(54.11)**	**Big Idea 4**

Chapter Overview

Behavioral biology is the study of what animals do, how they respond to internal and external changes, and their interactions with other organisms. Some behaviors are instinctive, while others are learned. This chapter explores what governs different behaviors, including details on evolutionary origins, physiology, genetics, development, and communication.

54.1 The Natural History of Behavior

Recall It

Behavior is necessary for animals to survive and reproduce. Behavior can be analyzed in terms of mechanisms and evolutionary origin. Some behaviors are innate or instinctive, meaning that no learning is required. The causes of innate behavior, including hormones, genes, and nerve cells, are studied by the field of ethology.

Essential Knowledge covered
3.E.1: Individuals can act on information and communicate it to others.

Review It

Define *behavior*.

List the four levels at which behavior can be analyzed.

Use It

As described in this section, geese will roll an egg back into its nest if the egg rolls out. What is the key stimulus that triggers behavior, and why is it beneficial to the goose?

Recall It

Neuroethology is the field that examines how neurobiology relates to behavior. Both the peripheral nervous system and central nervous system, as described in Chapter 43, are involved in sensing environmental change and transmitting the nerve impulses that drive specific reactions. Hormones and neurotransmitters can have a pronounced effect on behavior.

Review It

Provide an example of how the following molecules influence animal behavior:

Molecule	Example
Testosterone	
Serotonin	
Estrogen	

54.3 Behavioral Genetics

Recall It

While some behaviors are governed by the environment around us, other behaviors appear to have a genetic origin. Behavioral genetics is the study heritable behaviors. Many types of studies including artificial selection, hybrid, and twin studies have linked genes and behavior. Some behaviors appear to be controlled by a single gene.

Essential Knowledge covered
2.E.3: Timing and coordination of behavior are regulated by various mechanisms and are important in natural selection.

Review It

Using your textbook, provide an example of the following types of tests that have been performed to demonstrate behavior can be genetic.

Type of study performed	Example	Result
Twin study		
Artificial selection		
Molecular biology		

Use It

Compare and contrast the maternal behaviors of female mouse that has the *fosB* gene and those that lack *fosB* genes.

Recall It

When animals develop behaviors from previous experiences, this is called learning. Learning is possible only within the boundaries set by evolution. Habituation is the decrease in response to repeated stimuli that produce no positive or negative consequences. Associative learning is a change in behavior by association of two stimuli or of a behavior and a response. Associative learning was first described by Ivan Pavlov. Some animals may have innate predispositions toward forming certain associations and not others based on their evolutionary boundaries.

Essential Knowledge covered
2.E.3: Timing and coordination of behavior are regulated by various mechanisms and are important in natural selection.

Review It

Identify if the following is an example of habituation **(H)** or associate learning **(A)**:

Young birds become still when a large shadow flies overhead, but learn not to respond when leaves fall from the tree over their nests.

A bird learns that mimicking the sound "Polly wants a cracker" when a human is nearby will result in getting a cracker.

Birds do not eat insects that look like twigs.

Use It

If a pigeon is offered two seeds, one that is blue and one that is brown, and the blue one makes the pigeon sick, the pigeon can learn to avoid these in the future. However, the pigeon cannot associate food with sounds, so if brown poisoned seeds that look similar to the non-poisoned seeds are offered when a tone is played, the pigeon will eat the seeds. Explain why this might be.

Recall It

Young animals often form an attachment to other individuals or develop preferences that influence later behavior. Parent-offspring contact is important for animals learning social behaviors. Instinct and learning may interact as behavior develops. Animals may also have an innate genetic template that guides their learning as behavior develops, such as song development in birds. Studies on twins reveal a role for both genes and environment in human behavior.

Essential Knowledge covered
2.E.3: Timing and coordination of behavior are regulated by various mechanisms and are important in natural selection.

Review It

Which of the following statements are FALSE regarding animal learning?

    A.   Social attachments to other individuals at an early age may influence behavior later in life.

    B.   Parent-offspring contact is not necessary for learning social behavior.

    C.   White-crowned sparrows have a critical period of development they go through in which they need to learn how to sing.

    D.   Geese will only follow other geese when they are younger.

Use It

How do social interactions assist white-crowned sparrows learning how to sing?

Recall It

Cognitive behavior is the ability to process information and respond in a manner that suggests thinking. There is evidence that some nonhuman animals portray cognitive behaviors.

Review It

Using this section, describe an example of cognitive behavior that has been documented in nonhuman animals.

Recall It

Migration is the predictable back-and-forth movement of animals over long distances. In order for a population to migrate, the individuals must be capable of orientation and navigation. Animals orient themselves through tracking environmental stimuli such as celestial clues or by following Earth's magnetic field. Other animals navigate by following a route based on orientation and some sort of "map." The precise nature of the map in animals is not known.

Review It

List two ways birds can point themselves in the correct direction in order to migrate to the right location.

Why do monarchs head south from eastern North America to central Mexico each fall?

54.8 Animal Communication

Recall It

Communication enables for the exchange of information to pass between individuals and among groups. There are many types of signals involved in communication: visual, acoustic, chemical, electric, and vibrational. Sending messages through these means allows for successful courtship, reproduction, and exchange of information about food and predators.

Review It

Describe a mode of communication used by the following animals:

Animal	Mode of Communication
Bee	
Prairie dog	
Ants	
Lightening bugs	

Recall It

Behavioral ecology is the study of how natural selection shapes behavior. Natural selection favors optimal foraging strategies in which energy expenditure is minimized and reproductive success is maximized. Territorial behavior evolves if the benefits of holding a territory exceed the costs. Foraging behavior can directly influence energy intake and individual fitness.

Review It

Use the figure below to describe the optimal foraging behavior of the shore crab.

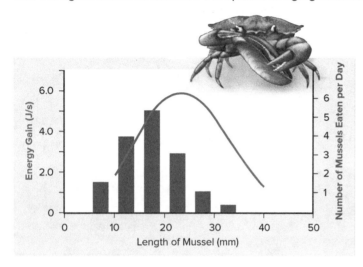

54.10 Reproductive Strategies and Sexual Selection

Recall It

Different reproductive strategies have evolved to maximize the reproductive success of a species. One sex may be choosier than the other, producing differences in morphology between sexes. These differences are called sexual dimorphisms. Intrasexual selection involves competition among members of the same sex for the chance to mate. Intersexual selection is one sex choosing a mate. There are direct benefits of being able to choose a mate, such as increased resource availability or parental care, and often indirect benefits, such as genetic quality of the mate. Mating systems include monogamy, polygyny, and polyandry; they are influenced by ecology and constrained by needs of offspring.

Review It

Define sexual dimorphism and provide an example.

Identify the mating systems based on the description below:

Mating System	Description
	One male mates with one female
	One male mates with more than one female
	One female mates with more than one male

54.11 Altruism

Recall It

Behavior that benefits another at the cost to the actor is called altruism. There are a number of explanations of why this behavior has evolved. Altruism may have arisen because mutual exchanges benefit both participants, and when a participant does not reciprocate, it would not receive future aid. Another thought is that kin selection increases the reproductive success of relatives and increased frequency of alleles shared by kin, so it is advantageous to be altruistic toward a relative as it increases an individual's inclusive fitness. Ants, bees, and wasps have haplodiploid reproduction, and therefore a high degree of gene sharing.

Essential Knowledge covered
4.B.3: Interactions between and within populations influence patterns of species distribution and abundance.

Review It

Name two theories have been developed to explain altruism.

List two factors that have contributed to the evolution of altruism.

Use It

Imagine you were snorkeling and witnessed a pod of dolphins chasing and catching fish. Sometimes, one would chase a fish in another's direction, so the other could catch the fish and eat. How could you determine if this behavior was reciprocal or kin selection?

Recall It

A society is a group of organisms of the same species that are organized in a cooperative manner. Insect societies form efficient colonies that contain specialized castes specialized to reproduce or to perform certain colony maintenance tasks. Vertebrate social systems are less rigidly organized. Vertebrate social systems tend to be influenced by food availability and predation.

Review It

The naked mole rat society is a major exception to the differences found between insect and vertebrate communities. In what ways are naked mole rats societies more like insect societies then vertebrate societies?

---

54 Chapter Review

Summarize It

The bobolink is a migratory bird that has an established summer range of the Midwestern United States. Scientists have recently seen that this bird is expanding its summer range. Where is it extending its range? What navigational tools might the bobolink use to change its ancestral migration route to and from this new range?

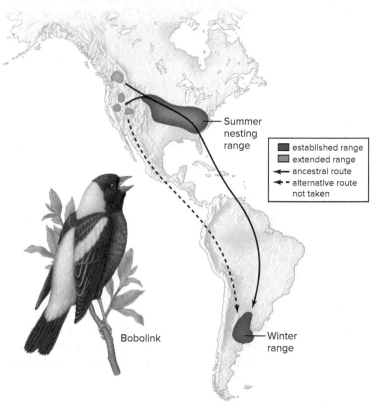

Summer nesting range

■ established range
■ extended range
← ancestral route
←- alternative route not taken

Bobolink

Winter range

Chapter 55: Ecology of Individuals and Populations

Essential Knowledge

2.D.1	All biological systems from cells and organisms to populations, communities, and ecosystems are affected by complex biotic and abiotic interactions involving exchange of matter and free energy. **(55.2, 55.3, 55.4)**	**Big Idea 2**
4.A.5	Communities are composed of populations of organisms that interact in complex ways. **(55.5, 55.6, 55.7)**	**Big Idea 4**
4.A.6	Interactions among living systems and their environment result in the movement of matter and energy. **(55.1, 55.7)**	**Big Idea 4**
4.C.3	The level of variation in a population affects population dynamics. **(55.1)**	**Big Idea 4**

Chapter Overview

Ecology is the study of how organisms interact with one another and their environment. Organismal growth and interactions are driven by both biotic and abiotic factors. Many species have developed unique life histories to maximize their biotic potential.

55.1 The Environmental Challenges

Recall It

The abiotic factors of temperature, water, sunlight, and soil type often determine what types of organisms can live in which environment. Most organisms are capable of responding to environmental changes and can cope with short-term changes in factors such as temperature and water availability. Natural selection leads to evolutionary adaptation to environmental conditions. Over evolutionary time, physiological, morphological, or behavioral adaptations evolve, shaped by the environment in which they live.

Essential Knowledge covered
4.A.6: Interactions among living systems and with their environment result in the movement of matter and energy.
4.C.3: The level of variation in a population affects population dynamics.

Review It

List three major environmental conditions that affect species distribution.

Describe two ways animals may adapt to their environment over time.

Use It

Using the graph below, explain how wolves and polar bears adapt to withstand differences in temperature throughout the year.

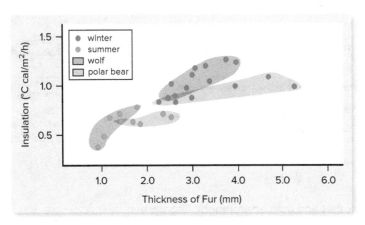

Some frogs have specialized glands which allow them to secrete a waxy substance to coat their skin. What advantage might this frog have in a climate experiencing drier conditions over frogs that do not have this ability?

55.2 Populations: Groups of a Single Species in One Place

Recall It

A population is a group of species that occur at one place at one time. A range is the population's geographic distribution. Most populations have limited ranges that expand or contract over time as the environment changes. Dispersal mechanisms may allow some species to expand their range by crossing barriers. Within a population, individuals may exhibit different spacing patterns: random, clumped, or uniform. Nonrandom distributions may reflect resource distributions or competition for resources. Networks of distinct populations that exchange individuals are called metapopulations. Metapopulations occur when populations are large and more connected, and may act as a buffer against extinction by permitting recolonization of vacant areas or marginal areas.

Essential Knowledge covered
2.D.1: All biological systems from cells and organisms to populations, communities and ecosystems are affected by complex biotic and abiotic interactions involving exchange of matter and free energy.

Review It

List three important characteristics of population ecology.

Identify the spacing patterns in the example populations described below:

Example	Spacing Pattern
A population where individuals do not interact sternly with one another and are not affected by non-uniform aspects of their habitat	
A population responds to uneven distribution of resources	
A population with competition for resources	

Draw an arrow indicating the movement of individuals between the two different metapopulations below:

Source   Sink

Use It

Penguins can be highly territorial. Given this information, what type of spacing pattern would you expect to see in their population, and why?

Describe two important implications of metapopulations on the potential range of a species.

Recall It

Demography is the quantitative study of populations. Populations with short generation times or many reproductive females exhibit rapid population growth. The age structure of a population also affects the growth of a population. Age structure is determined by the numbers of individuals in different age groups, called cohorts. Every age cohort also has a characteristic fecundity and death rate. Life tables are constructed to show the probability of survival and reproduction through a cohort's life span. Survivorship curves demonstrate how survival probability changes with age. In some populations, survivorship is high until old age, whereas in others, survivorship is lowest among the youngest individuals. Survivorship curves are conventionally placed into three categories: Type I, Type II, and Type III.

Essential Knowledge covered
2.D.1: All biological systems from cells and organisms to populations, communities and ecosystems are affected by complex biotic and abiotic interactions involving exchange of matter and free energy.

Review It

Identify if the statement pertains to survivorship curve **I**, **II**, or **III**:

Humans follow this type of survivorship curve.

Mortality rates are equal at any age.

Individuals produce many offspring but most do not survive to reproductive age.

Mortality rates rise steeply later in life.

Use It

Describe the relationship between body size and generation time. Which organism would you expect to see have a faster growing population: a housefly or a bear?

Recall It

Traits that maximize the number of surviving offspring in the next generation are favored by natural selection. These traits are affected by how long an individual lives, and how many offspring are produced. Differences in life histories encompass many trade-offs. When reproductive cost is high, fitness can be maximized by deferring reproduction, or by producing a few large-sized young that have a greater chance of survival. Resources allocated toward current reproduction cannot be used to enhance survival and future reproduction. A trade-off exists between number of offspring and investment per offspring. Reproductive events per lifetime represent an additional trade-off. Some organisms follow a life history called semelparity, in which reproduction occurs in a single large event. Iteroparity occurs when individuals produce offspring several times over many seasons. Reproduction may occur later in longer-lived species compared with short-lived species.

Essential Knowledge covered
2.D.1: All biological systems from cells and organisms to populations, communities and ecosystems are affected by complex biotic and abiotic interactions involving exchange of matter and free energy.

Review It

In ecological terms, describe the phrase "the cost of reproduction."

Identify if the statement pertains to iteroparity **(I)** or semelparity **(S)**:

Usually found in short-lived species.

Results in the production of several offspring over many seasons.

Species with this reproductive strategy do not use up their energy during reproduction.

All reproductive resources are focused on one single reproductive event, and then the organism often dies.

Use It

Life histories are all about trade-offs. Describe the trade-off illustrated in the graph below showing the size of nestlings (baby birds) related to clutch size (the number of eggs) of in the bird called the great tit.

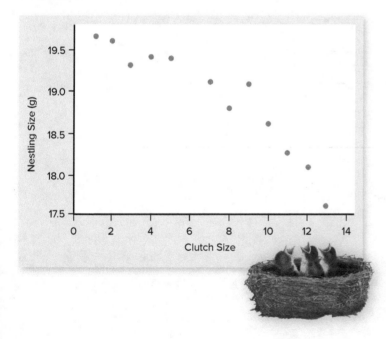

In the example above, what reproductive strategy do you think is better for this bird?

Recall It

There are two types of growth models commonly seen in population ecology: exponential growth and logistic growth. Exponential growth occurs when a population is not limited by resources or by other species. Logistic growth is observed as a population reaches its carrying capacity. A carrying capacity is the maximum number of individuals that an environment can support and is where population growth stabilizes. In some cases, a population may overshoot and then drops back to the carrying capacity.

Essential Knowledge covered
4.A.5: Communities are composed of populations of organisms that interact in complex ways.

Review It

Provide a definition for the following population growth vocabulary:

Vocabulary	Definition
Exponential growth	
Birth rate (*b*)	
Logistic growth	
Biotic potential	
Sigmoidal growth	
Carrying capacity (*K*)	
Death rate (*d*)	

Use It

Use the graph below to answer the following questions:

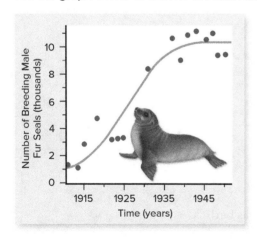

What is the carrying capacity of the number of breeding male fur seals in this population?

What year does the fur seal reach its carrying capacity?

Describe the growth model of the fur seal.

If the fur seal growth was not limited by the amount of resources it had available (example: fish it could catch and eat), what type of growth pattern would you expect to see?

55.6 Factors That Regulate Populations

Recall It

Many factors may regulate population size. These are categorized as density-dependent factors and density-independent factors. Density-dependent factors are biotic, and occur when reproduction and survival are affected by population size. These include factors like increased competition and disease. Density-independent factors are abiotic, and include environmental disruptions and catastrophes. As the name implies, density-independent factors are not related to population size. In some cases, population size is cyclic because of the interaction of factors such as food supply and predation. Populations have been historical categorized as K-selected and r-selected, depending on their life history adaptations. K-selected species are adapted to thrive at carrying capacity, while r-selected species have high reproductive rates and tend to exist below carrying capacity.

Essential Knowledge covered
4.A.5: Communities are composed of populations of organisms that interact in complex ways.

Review It

Define *density-independent* and *density-dependent* factors and provide an example of each:

	Definition	Example
Density-independent factor		
Density-dependent factor		

Identify if the following statements pertain to *K*-selected **(K)** or *r*-selected **(r)** species:

Produce fewer, stronger offspring.

Produce a maximum number of offspring and thrive when resources are not limiting.

Thrive near their carrying capacity.

Use It

Construct a graph to help explain how interactions between herbivores, edible plants, and predators can lead to cyclical patterns of population increase and decrease.

Are all density-dependent factors negatively related to population size? Explain your answer.

Recall It

The human population has experienced exponential growth over the past 300 years. Technology and other innovations have simultaneously increased the carrying capacity and decreased mortality. Population pyramids show that populations with many young individuals are likely to experience high growth rates as these individuals reach reproductive age. The global human population is unevenly distributed with rapid growth occurring in developing countries. However, resource consumption rates in the developed world are higher than that of the undeveloped areas. Finding a path to a sustainable future requires figuring out ways limit both population growth and per capita resource consumption. Even at lower growth rates, the number of individuals on the planet is likely to plateau at 7 to 10 billion.

Essential Knowledge covered
4.A.5: Communities are composed of populations of organisms that interact in complex ways.
4.A.6: Interactions among living systems and with their environment result in the movement of matter and energy.

Review It

List three reasons why humans have experienced exponential growth.

Define population pyramid.

What is an ecological footprint?

Circle the country with the greatest ecological footprint.

United States   India   Nigeria

Use It

Draw a diagram of a population pyramid showing a stable population and a population experiencing rapid growth. Which has more individuals of the reproductive age?

Explain why lowering a population's ecological footprint is an important step in sustaining life on Earth.

Summarize It

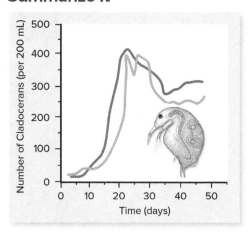

1. Using the graph above, determine the carrying capacity for these cladoceran populations. Describe what happens to the populations when it first exceeds its carrying capacity.

2. Age of first reproduction in animal populations is often correlated with life span: longer-lived species reproduce later in their lives, while shorter-lived species reproduce sooner. Construct an explanation of how this difference in reproductive strategy may have evolved.

Chapter 56: Community Ecology

Essential Knowledge

2.D.1	All biological systems from cells and organisms to populations, communities, and ecosystems are affected by complex biotic and abiotic interactions involving exchange of matter and free energy. **(56.1, 56.3, 56.4)**	**Big Idea 2**
2.D.3	Biological systems are affected by disruptions to their dynamic homeostasis. **(56.5)**	**Big Idea 2**
2.E.3	Timing and coordination of behavior are regulated by various mechanisms and are important in natural selection. **(56.2)**	**Big Idea 2**
4.A.5	Communities are composed of populations of organisms that interact in complex ways. **(56.1, 56.4)**	**Big Idea 4**
4.B.3	Interactions between and within populations influence patterns of species distribution and abundance. **(56.3, 56.4)**	**Big Idea 4**
4.C.4	The diversity of species within an ecosystem may influence the stability of the ecosystem. **(56.4)**	**Big Idea 4**

Chapter Overview

Community ecology is the study of how species living together in one place interact. These interactions include both competition and cooperation. Community ecology also studies how communities change over time.

56.1 Biological Communities: Species Living Together

Recall It

A community is defined as all the species that live together in any particular environment. Communities can be measured through species richness or through primary productivity. Individual species generally respond independently to environmental conditions. Community composition gradually changes over space and time. However, in locations called ecotones, where conditions rapidly change, species composition may change greatly over short distances.

> **Essential Knowledge covered**
> *2.D.1: All biological systems from cells and organisms to populations, communities and ecosystems are affected by complex biotic and abiotic interactions involving exchange of matter and free energy.*
> *4.A.5: Communities are composed of populations of organisms that interact in complex ways.*

Review It

List three ways communities may experience change in composition.

Give an example of an ecotone.

Use It

Do the different species in a community always change in a synchronous pattern over time or space? Explain your answer.

56.2 The Ecological Niche Concept

Recall It

The ways in which an organism uses the resources of its environment is called its niche. A niche can be fundamental or realized. The fundamental niche is the entire niche a species is capable of using, but competition between species creates what is known as a realized niche. Sometimes when species compete for resources, competitive exclusion can occur, in which the more efficient species will out compete the less efficient species. It is possible for two or more species to coexist in a niche if they partition the use of resources. Partitioning resources leads to reduce competition.

Essential Knowledge covered
2.E.3: Timing and coordination of behavior are regulated by various mechanisms and are important in natural selection.

Review It

Provide a definition for the following ecological vocabulary words:

Vocabulary	Definition
Partitioning	
Fundamental niche	
Interspecific competition	
Niche	
Competitive exclusion	
Realized niche	

Use It

Owls and hawks both live in forests and both eat the same types of small rodents. How are they able to occupy the same habitat?

Recall It

Predation occurs when one organism consumes another, and includes both carnivore and herbivore activity. The interaction of predator and prey populations are complex; both are needed to help regulate the other's population. Predator-prey relationships may lead to an evolutionary "arms race," in which the predator and prey co-evolve better defenses and better means of eluding capture. For example, an evolutionary arms race exists between some plants which produce toxic chemicals and herbivores which have evolved the ability to withstand these chemicals. Animal adaptations include chemical defenses and defensive coloration such as warning coloration or camouflage. Mimicry allows one species to capitalize on defensive strategies of another. Batesian mimicry occurs when a species that is edible or nontoxic evolves warning coloration similar to that of an inedible or poisonous species. In Müllerian mimicry two species that are both toxic evolve similar warning coloration.

Essential Knowledge covered
2.D.1: All biological systems from cells and organisms to populations, communities and ecosystems are affected by complex biotic and abiotic interactions involving exchange of matter and free energy.
4.B.3: Interactions between and within populations influence patterns of species distribution and abundance.

Review It

Identify the following defensive adaptations based on the examples provided below:

Example	Adaptation
A scorpion sprays venom to protect itself.	
Monarch butterflies are bright orange to warn predators they contain poisonous toxins.	
Another nonpoisonous butterfly is also bright orange.	
An inchworm appears to closely resemble a stick.	
Three poisonous butterflies are all brightly colored orange.	

Place an up arrow (↑) or a down arrow (↓) for what you would expect for the predator populations given the following scenarios.

A predator exterminates all of its prey.

A prey population increases dramatically.

A human eliminates prey from an area.

Use It

Tiger moths have evolved the ability to jam the sonar, or echolocation waves, that brown bats use to hunt. What can you infer about the relationship of these two species?

Describe what might happen to a population of white-tailed deer if humans remove all the large carnivorous predators from the area.

56.4 The Many Types of Species Interactions

Recall It

Symbiosis is the close, long-term relationship between two different species. The major types of symbiosis are mutualism, parasitism, and commensalism. Mutualism benefits both species. In parasitism, one species benefits and the other is harmed. During commensalisms, one species benefits but the other is neither helped nor harmed by the relationship. Because many processes may occur simultaneously, species may affect one another not only through direct interactions but also through their effects on other species in the community. Keystone species are those that maintain a more diverse community by reducing competition between species or by altering the environment to create new habitats.

Essential Knowledge covered
2.D.1: All biological systems from cells and organisms to populations, communities and ecosystems are affected by complex biotic and abiotic interactions involving exchange of matter and free energy.
4.A.5: Communities are composed of populations of organisms that interact in complex ways.
4.B.3: Interactions between and within populations influence patterns of species distribution and abundance.
4.C.4: The diversity of species within an ecosystem may influence the stability of the ecosystem.

Review It

Identify the symbiotic relationship described in the examples below:

Example	Symbiosis
A hummingbird feeds on a flower and spreads the flower's pollen.	
A wasp lays eggs in a living host, and the larvae feed from the body of the host, slowly killing the host.	
A barnacle settles on a whale and is transported from place to place.	

Define keystone species.

Use It

A biodiversity survey was performed in two intertidal zones. One site (Site A) had a small population of sea stars, while the other site (Site B) did not have any sea stars. Based on the data below, what can you determine about the nature of the sea star?

	Site A	Site B
Number of species counted	24	9

Using the relationship between an oxpecker bird and an antelope, describe under what conditions their relationship would be classified as (A) commensal, (B) mutualistic, and (C) parasitic.

56.5 Ecological Succession, Disturbance, and Species Richness

Recall It

Succession is the sequential change of species composition in a community. Primary succession occurs at the site of a barren, lifeless substrate. Secondary succession occurs after an existing community experiences an ecological disturbance. Ecological disturbances include forest fires, clearing cutting, or other events. Disturbances play an important role in structuring communities. Community composition changes because of local and global disturbances that "reset" succession. Intermediate levels of such disturbance may maximize species richness in two ways: by creating a patchwork of different habitats harboring different species, and by preventing communities from reaching the final stage of succession, which may be dominated by only a few, competitively superior species.

Essential Knowledge covered
2.D.3: Biological systems are affected by disruptions to their dynamic homeostasis.

Review It

What is the major difference between primary and secondary succession?

Use It

The intermediate disturbance hypothesis states that a moderate amount of disturbance in an ecosystem can actually increase species diversity. How is this possible?

56 Chapter Review

Summarize It

1. The graph above shows the beak depths in two species of seed-eating finches, *G. fuliginosa* (blue) and *G. fortis* (red), when they are alone on islands Los Hermanos and Daphne Major, respectively. When they cohabitate on San Cristóbal and Floreana, differences are observed in their beak size. What are these differences and what might account for the change?

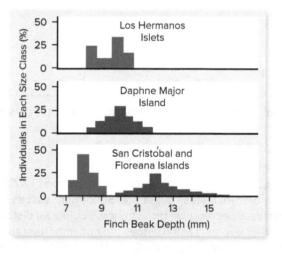

2. Alligators excavate deep holes at the bottoms of lakes in areas that are prone to drought. During years that are very dry, these waterholes remain even after the water in the lakes dry up. If the alligators were hunted to extinction, what might happen to other organisms living in that entire ecosystem surrounding that lake?

Chapter 57: Dynamics of Ecosystems

Essential Knowledge

2.A.1	All living systems require constant input of free energy. **(57.2, 57.3)**	Big Idea 2
2.A.3	Organisms must exchange matter with the environment to grow, reproduce and maintain organization. **(57.1)**	Big Idea 2
2.D.1	All biological systems from cells and organisms to populations, communities and ecosystems are affected by complex biotic and abiotic interactions involving exchange of matter and free energy. **(57.1, 57.2)**	Big Idea 2
4.A.5	Communities are composed of populations of organisms that interact in complex ways. **(57.4)**	Big Idea 4
4.A.6	Interactions among living systems and with their environment result in the movement of matter and energy. **(57.2, 57.3)**	Big Idea 4
4.B.3	Interactions between and within populations influence patterns of species distribution and abundance. **(57.3)**	Big Idea 4
4.B.4	Distribution of local and global ecosystems changes over time. **(57.1)**	Big Idea 4

Chapter Overview

The movement of energy and matter within an ecosystem is driven by the interactions between species. As you learned in Chapter 7, all organisms require energy for cellular respiration, and the paths of energy transfer between organisms create what are known as trophic levels within an ecosystem. The number of different species found in an ecosystem are important in the overall health and stability of that ecosystem.

57.1 Biogeochemical Cycles

Recall It

An ecosystem includes both the abiotic and biotic components of the environment. These components are dynamic. Biogeochemical cycles describe how chemical elements cycle though an ecosystem. This chapter describes four biogeochemical cycles: (1) the carbon cycle, (2) the water cycle, (3) the nitrogen cycle, and (4) the phosphorous cycle. These four abiotic components are all critical in supporting the biotic factors in an ecosystem. These natural cycles can be disrupted by human activities, such as logging, agriculture, and industrial practices, which can lead to instability in the ecosystem.

Essential Knowledge covered
2.A.3: Organisms must exchange matter with the environment to grow, reproduce, and maintain organization.
2.D.1: All biological systems from cells and organisms to populations, communities and ecosystems are affected by complex biotic and abiotic interactions involving exchange of matter and free energy.
4.B.4: Distribution of local and global ecosystems change over time.

Review It

Define the following vocabulary terms related to biogeochemical cycles:

Vocabulary	Definition
Aquifers	
Nitrogen fixation	
Nitrification	
Denitrification	

Use It

All life on Earth requires water. Around 50% of the population of the United States obtain their drinking water from groundwater. How does the groundwater get there and how is it recharged? Use a drawing to help illustrate your answer.

Describe how deforesting a tropical forest may lead to changes in the water cycle.

Compare and contrast the nitrogen and phosphorous cycles.

Recall It

As you will recall from Chapter 6, energy exists in forms such as light, chemical bonds, motion, and heat, and is governed by the laws of thermodynamics. In ecosystems, energy flows through trophic levels. Organic compounds are synthesized by autotrophs and are utilized by both autotrophs and heterotrophs. Each level, or organism, energy passes through is termed a trophic level. The sequential progression of energy through the trophic levels is called a food chain. The base trophic level includes the primary producers; herbivores that consume primary producers are the next level. They in turn are eaten by primary carnivores, which may be consumed by secondary carnivores. Detritivores feed on waste and the remains of dead organisms. As energy moves through each trophic level, very little (approximately 10%) remains from the preceding trophic level. The exponential decline of energy between trophic levels limits the length of food chains and the numbers of top carnivores that can be supported. Ecological pyramids can be formed from food chains to show the energy flow, biomass, or numbers of organisms in each trophic level. It is possible for an ecological pyramid to be inverted if at least one trophic level has a greater biomass or more organisms than the level below it.

Essential Knowledge covered
2.A.1: All living systems require constant input of free energy.
2.D.1: All biological systems from cells and organisms to populations, communities and ecosystems are affected by complex biotic and abiotic interactions involving exchange of matter and free energy.
4.A.6: Interactions among living systems and with their environment result in the movement of matter and energy.

Review It

Identify the four tropic levels in the food chain below. Which organisms are autotrophs? Which organisms are heterotrophs?

berry bushes   →   song birds   →   weasel   →   fox

Define detritivore.

Use It

Use the ecological biomass pyramid to answer the following questions:

Construct a food chain using the three trophic levels illustrated in this pyramid.

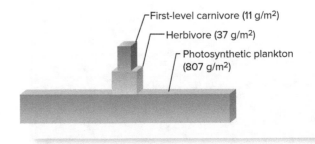

First-level carnivore (11 g/m²)

Herbivore (37 g/m²)

Photosynthetic plankton (807 g/m²)

Imagine another herbivore, too big for the carnivore to eat, was introduced into the system illustrated on p. 310. What would happen to the first-level carnivores in this pyramid if the photosynthetic plankton biomass was suddenly and dramatically reduced?

Under what conditions you would expect to see the second trophic level have a greater biomass (bigger bar) than the primary trophic level?

57.3 Trophic-Level Interactions

Recall It

When a change exerted at an upper trophic level affects a lower trophic level, this is known as a trophic cascade. The changes are considered top-down effects. An example of a top-down effect can be seen when a removal of carnivores causes an increase in the abundance of species in lower trophic levels. Bottom-up effects can also occur when changes to primary producers affect higher trophic levels.

Essential Knowledge covered
2.A.1: All living systems require constant input of free energy.
4.A.6: Interactions among living systems and with their environment results in the movement of matter and energy.
4.B.3: Interactions between and within populations influence patterns of species distribution and abundance.

Review It

Using the following food chain, explain what organisms would need to be affected to produce either top-down or bottom-up effects.

Kelp → Sea urchins → Sea otters → Orcas

Use It

Explain why the removal of orcas in kelp forest ecosystems leads to the reduction of sea urchins and the increase in kelp.

Recall It

There is evidence that a community with higher species richness has less year-to-year variation in biomass and greater resistance to drought and invasion by nonnative species. The number of species found in an ecosystem can be affected by primary production, habitat heterogeneity, and climatic factors. The higher diversity of tropical regions may be the result of many factors, including long evolutionary time, higher productivity from increased sunlight, less seasonal variation, greater predation that reduces competition, and/or spatial heterogeneity.

Essential Knowledge covered
4.A.5: Communities are composed of populations of organisms that interact in complex ways.

Review It

Define species richness.

List three factors that influence species richness.

Use It

Why is species richness important?

57.5 Island Biogeography

Recall It

Island biogeography is the study of how species colonize small islands. The island biogeography equilibrium model says that colonization and extinction rates equilibrate at some point, and this rate determines the number of species an island can support. The equilibrium shifts depending on how far the island is from the mainland and the size of the island.

Review It

Would you expect an island near a mainland to have a greater rate of colonization than an island that is far from the mainland? Why or why not?

Which island would have a greater extinction rate: a small island or a large island? Why?

---

57 Chapter Review

Summarize It

1. The carbon cycle is an outwardly straightforward process. In what way are humans affecting the natural flow of carbon through the ecosystem?

2. Predict how the introduction of an insectivorous fish would change the population of the community of the pond illustrated in the graph below.

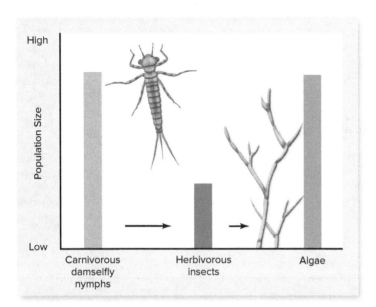

Chapter 58: The Biosphere

Essential Knowledge

4.A.6	Interactions among living systems and with their environment result in the movement of matter and energy. **(58.5, 58.6)**	**Big Idea 4**
4.C.2	Environmental factors influence the expression of the genotype in an organism. **(58.6)**	**Big Idea 4**

Chapter Overview

The biosphere encompasses all the ecosystems of Earth. These include eight major biomes, which have distinct climatic conditions, vegetation, and animal communities. Unfortunately, human activities are creating adverse change in many of these biomes, by adding pollution, depleting resources, and accelerating climate change.

58.1 Ecosystem Effects of Sun, Wind, and Water

Recall It

Variation in solar radiation and patterns of atmospheric and oceanic circulation heavily influence the global patterns of life on Earth. The curvature in the paths winds take as they move is called the Coriolis Effect, and is caused by Earth's rotation. Global currents, in turn, are mostly driven by the winds. Ecosystems can also be affected by local and regional changes in solar radiation, as well as air circulation, water circulation, and changes in soil types. Changes in altitude can also affect temperature.

Review It

Identify the following ecosystem vocabulary terms based on their definition:

Definition	Vocabulary
Seasonal massive rain that moves from the oceans onto land.	
The factor that drives the curved path of winds and surface currents on Earth.	
One side of a mountain is very dry, while the other that faces the ocean is moist.	
A giant closed curve of moving ocean water.	

Explain the difference between weather and climate.

Recall It

Biomes are the major ecosystems found on Earth. There are eight major biomes: (1) tropical rainforest, (2) savanna, (3) desert, (4) temperate grassland, (5) temperate deciduous forest, (6) temperate evergreen forest, (7) taiga, and (8) tundra. Each biome has a unique vegetation structure, climate, and different communities of secondary consumers.

Review It

Describe the climate and vegetation of the following biomes:

Biome	Climate	Vegetation Type
Tundra		
Desert		
Temperate grasslands		
Savanna		
Taiga		
Temperate deciduous forest		
Temperate evergreen forest		
Tropical rain forest		

58.3 Freshwater Habitats

Recall It

Freshwater habitats cover only 2% of Earth's surface but represent some of the most productive ecosystems on Earth. Freshwater habitats include ponds, lakes, rivers, and different types of wetlands. Dissolved oxygen concentrations are important for freshwater organisms. Light is another important property. The areas that receive enough light for photosynthesis to occur are called photic zones. In temperate climates, when the top of the surface of the lake cools, it sinks, in what is known as the fall turnover. In the spring, the warm water rises and the cold water sinks, creating thermal stratification. Lakes are classified as eutrophic or oligotrophic, depending on the amount of nutrients and dissolved oxygen they contain.

Review It

List two abiotic factors that are important to organisms living in freshwater habitats.

Explain the difference between an oligotrophic and eutrophic lake.

58.4 Marine Habitats

Recall It

The ocean is divided into several zones based on nutrient availability, light, tidal movement, and depth. Phytoplankton are the dominant primary producers in most of the open ocean. Many coastal habitats are rich in rooted plants and algae as well. Upwelling occurs when tradewinds bring cold, nutrient-rich waters to the surface from below, increasing concentrations of nutrients and oxygen at the surface. El Niño events weaken tradewinds, restricting upwelling to surface waters, which causes a drop in the amount of phytoplankton productivity. The deep sea is the single largest habitat, and contains hydrothermal vent communities occur which possess unique chemosynthetic organisms.

Review It

Draw a diagram which explains how upwelling occurs in the ocean.

58.5 Human Impacts on the Biosphere: Pollution and Resource Depletion

Recall It

Human activities can cause many adverse changes in the ecosystem. Freshwater habitats are threatened by pollution and resource depletion. Different types of pollution include point source pollution, diffuse pollution, and acid precipitation. Terrestrial habitats are often threatened by deforestation, which leads to loss of habitat, disruption of the water cycle, and loss of nutrients. Acid rain can negatively affect both terrestrial and aquatic habitats. Habitat fragmentation and human interactions with the environment are leading to increases in zoonotic diseases. CFC emissions lead to ozone depletion in the atmosphere. Overfishing is a major problem that plagues the oceans.

Essential Knowledge covered
4.A.6: Interactions among living systems and with their environment result in the movement of matter and energy.

Review It

Define the following ecological threats:

Threat	Definition
Ozone depletion	
Habitat Loss	
Overfishing	
Pollution	
Acid Precipitation	
Eutrophication	
Deforestation	

Use It

Describe the possible outcomes of overfishing the oceans.

What causes the ozone layer to be depleted, and what negative consequences can this have on organism?

Recall It

Carbon dioxide (CO_2) levels in the atmosphere are important for regulating Earth's temperature. Known as a greenhouse gas, CO_2 allows solar radiation to pass through the atmosphere but prevents heat from leaving Earth, creating warmer conditions. If temperatures change rapidly, natural selection cannot occur rapidly enough to prevent many species from becoming extinct. Rapid global warming may also cause changes in sea levels, increase frequency of extreme climatic events, impact agriculture, and increase the range of tropical diseases.

Essential Knowledge covered
4.A.6: Interactions among living systems and with their environment result in the movement of matter and energy.
4.C.2: Environmental factors influence the expression of the genotype in an organism.

Review It

Identify if the following statements regarding climate change are true or false **(T/F)**.

There is evidence that the global temperature is rising.

Global warming poses no threat to organisms that live in water.

Global warming only has negative effects on agriculture.

Greenhouse gases hinder the escape of heat into space.

Global warming may cause malaria to spread to the United States.

Use It

What evidence is there that support that Earth's climate is warming? Why should we be concerned?

Summarize It

Describe three ways animals may respond to global climate change.

Chapter 59: Conservation Biology

Essential Knowledge

1.A.2	Natural selection acts on phenotypic variations in populations. **(59.1)**	Big Idea 1
1.C.1	Speciation and extinction have occurred throughout the Earth's history. **(59.3)**	Big Idea 1
4.B.4	Distribution of local and global ecosystems changes over time. **(59.3)**	Big Idea 4
4.C.3	The level of variation in a population affects population dynamics. **(59.3)**	Big Idea 4
4.C.4	The diversity of species within an ecosystem may influence the stability of the ecosystem. **(59.3)**	Big Idea 4

Chapter Overview

Global extinction rates have accelerated alarmingly. Conservation biology is the field devoted to learning the best methods for the preservation of species, communities, and ecosystems. Understanding what causes populations to go extinct is an important aspect of conservation biology.

59.1 Overview of the Biodiversity Crisis

Recall It

While extinction events have occurred through geologic history, recent losses of biodiversity are directly related to human activities. When humans arrived in North America after the last Ice Age, at least 75% of large mammals became extinct. Since 1600, 85 mammal and 113 bird species have gone extinct. The majority of historical extinctions have occurred within the last 150 years and have occurred on islands. Endemic species, those found in one restricted range, are especially threatened. Species hotspots are areas with many endemic species. Many hotspots are the site of large human population growth and face the highest rates of extinction.

Essential Knowledge covered
1.A.2: Natural selection acts on phenotypic variations in populations.

Review It

Define the following conservation biology vocabulary terms:

Vocabulary	Definition
Extinction	
Hotspot	
Endemic species	

Use It

Describe how prehistoric humans changed local populations of other species.

Why are biodiversity hotspots facing a greater threat of extinction than other locations?

59.2 The Value of Biodiversity

Recall It

Biodiversity is valuable to humans in a number of ways. There is a direct economic value associated with maintenance of biodiversity. Many commercial products are derived from a variety of plants and animals. There is also an indirect economic value in the form of ecosystem services. Ecosystem services include chemical processing, buffering against storm surges and droughts, soil preservation, and pollution absorption. Biodiversity also provides ethical and aesthetic value.

Review It

List four ways biodiversity is valuable to humanity.

Give an example of an "ecosystem service."

Recall It

There are many factors that contribute to extinction. These include habitat loss, overexploitation, and species introduction, as well unknown and combined reasons. Habitat loss can occur through destruction, pollution, disruption, or fragmentation. As habitats become more fragmented, the proportion of the remaining habitat that occurs on the boundary or edge increases exposing species to parasites, nonnative species, and predators. Overexploitation may occur through unregulated hunting and harvesting. Introduced or invasive species result in large and often negative changes to a community, usually because there are few impediments to the growth of the non-native species. Extinction of one species may trigger extinction cascades, either top-down or bottom-up, through the trophic levels. The extinction of a keystone species may increase competition and greatly alter ecosystem structure and function. Catastrophes, lack of mates, and loss of genetic variability all make reduced populations more likely to become extinct.

Essential Knowledge covered
1.C.1: Speciation and extinction have occurred throughout the Earth's history.
4.B.4: Distribution of local and global ecosystems change over time.
4.C.3: The level of variation in a population affects population dynamics.
4.C.4: The diversity of species within an ecosystem may influence the stability of the ecosystem.

Review It

Name three factors that might cause a species to go extinct.

List two ways an introduced species may threaten a native species or habitat.

Use It

Flying foxes are fruit-eating bats that pollinate many plants and disperse many seeds. Their populations are currently in decline because humans are hunting them and their habitats are being destroyed. What effects might the extinction of these flying foxes have on their habitats and communities?

The greater prairie chicken is found in the midwestern United States and has a flamboyant mating ritual. The population of prairie chickens has varied greatly throughout the last hundred years or so, facing many local extinctions. A prairie chicken sanctuary was erected to try to save the dwindling populations in the state of Illinois, but population rates kept dropping and egg-hatching rates plummeted as well. Why would this small population of prairie chickens face problems in reproductive success, even when protected from outside harm?

How might humans accidently spread a nonnative species to a new environment?

59.4 Approaches for Preserving Endangered Species and Ecosystems

Recall It

If the cause of a species's decline is known, scientists may be able to design a recovery plan. Plans are often expensive to implement, but may stave off even more expense ecological disturbances if the species is lost permanently. Conservationists might attempt to recreate a habitat by reintroducing plants and animals to an ecosystem, or to remove non-native or introduced species. Other conservation efforts include cleaning up chemical contaminants from a polluted ecosystem. Sometimes conservation efforts include starting a captive breeding program to establish a larger population of animals to re-introduce into a restored habitat.

Review It

Determine if the following statements regarding conservation biology are true or false **(T/F)**:

Extinction is less likely if species conservation is coupled with habitat restoration.

Introducing a new species to control a previously introduced species is common and effective conservation practice.

Captive breeding programs have been used to save some species from extinction.

Habitats degraded by chemical pollution cannot ever be restored.

Almost all conservation program solutions are inexpensive and effective.

Summarize It

TABLE	Recorded Extinctions Since 1600					
	RECORDED EXTINCTIONS				**Approximate Number of Species**	**Percent of Taxon Extinct**
Taxon	**Mainland**	**Island**	**Ocean**	**Total**		
Mammals	30	51	4	85	5,000	1.7
Birds	21	92	0	113	10,000	1.1
Reptiles	1	20	0	21	10,000	0.2
Fish	22	1	0	23	31,000	0.1
Invertebrates*	49	48	1	98	1,000,000+	0.01
Flowering plants	245	139	0	384	250,000	0.2

*Number of extinct invertebrates is probably greatly underestimated due to lack of knowledge for many species (other groups are probably underestimated to a lesser extent for the same reason).

1. The table above shows the number of species that have gone extinct since 1600. Use this table to answer the following questions:

   a. Where have the majority of mammal, bird, and reptile extinctions occurred? Why do you think this is?

   b. Why is the "Percent of Taxon Extinct" so low for invertebrates compared to other taxon? Are invertebrates better at escaping causes of extinction?

   c. Which location has the largest loss of fish species? Why do you think this is?

2. If you were a conservation biologist and you needed to design a captive breeding program to save a dwindling species, what factors would you consider to ensure that the program was successful?